AERODYNAMIC
DEFINITIONS

It is the first book in a large and special series of books, dedicated to motorsport in general; it will cover aerodynamics, suspension, engines, dynamics, etc. ..... Everything you need to learn how to design a full car.

The aim of this series is also to say that I would like to teach again in a university.

I hope that this series will be a success and that I will be able to transmit all my knowledge and all my experience.

@TimoteoBriet

DENSITY / PRESSURE

The density of a group of particles is defined as the mass of the group of particles per unit volume. In other words, that is the amount of particles mass that we can collect with a given unit container. When more particles fit in the container, the higher the density of the group will be.

Mathematically, we can express it in the following way:

$$\rho = \frac{m}{V}$$

"m" is the mass of the group and "V" is the total volume; the units density, in the S.I., are Kg / m³. In fact, the molecules of the air cannot be over than a certain distance nor closer than a certain distance. Both limits mark out the compressibility of the air. The air is composite by lot molecules:

In 10^{-9} mm³ there are $3 * 10^7$ molecules (in normal conditions). This volume is named V*.

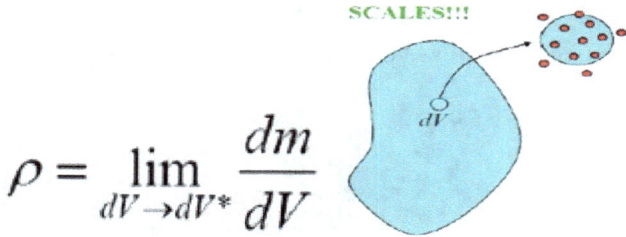

$$\rho = \lim_{dV \to dV^*} \frac{dm}{dV}$$

In this volume little, occur strange things:

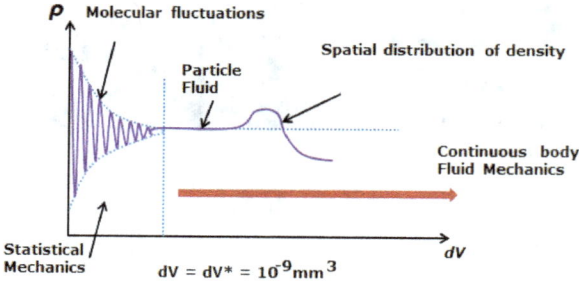

In aerodynamic terms, the higher is the density of the air that the car faces, the higher the *downforce* will be (we will see that in another chapter). The *drag* will be higher as well. In order to increase the air density, we can do as follows:

- Increasing the total pressure.

- Reducing the temperature.

But if the density change, the engine power change too; in fact, is possible to know the variation of % oxygen in air against temperature; also about the density (negative is reduction):

% Reduction Oxygen in Air	
Temperature (ºC)	
50	-11
30	-3,5
20	-2,5
10	-1,2
0	-0,6
-10	-0,2

% Reduction Air Density	
Temperature (ºC)	
50	-3,9
30	-1,55
20	-1,35
10	-0,55
0	-0,35
-10	-0,05

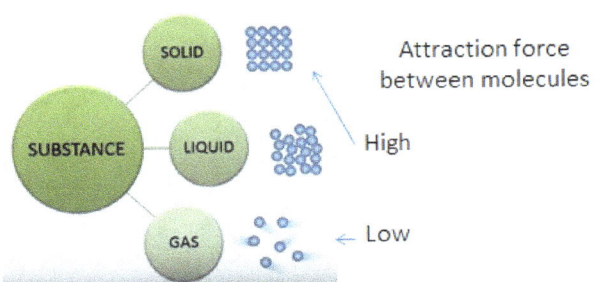

So the variation in engine power and variation downforce or drag, against air temperature variation, are different; that is very important.

Calculating the air density is not easy. We use indirect methods based in other parameters variations to know the variation of the air density.

These expressions are called equations of state. To do this, we also need a "base" air density in order to know its variations. Consequently, we need the standard atmosphere definition in order to determine its values:

Height (m)	Temperature (°C)	Pressure (HPa)	Density (kg/m³)
0	15	1013.2	1.23
500	11.8	955	1.17
1000	8.5	899	1.11

We can rely on the following table in order to know the "standard" values:

Air is a Gas. 78% Nitrogen, 21% Oxygen, traces H_2O, CO_2, Ar, ..			
Property	**Dimensions**	**Value (SLS*)**	
		Metric	Imperial
Mass, Volume Density (r)	mass/volume	1.229 kg/m³	.00237 slug/ft³
Specific Volume (v)	volume/mass	.814 m³/kg	422 ft³/slug
Pressure (p)	force/area	101.3 kN/m²	14.7 lb/in²
Temperature (T)	degrees	15 °C	59 °F
Viscosity (mu)	force-time/area	1.73×10^{-5} N-s/m²	3.62×10^{-7} lb-s/ft²
* Sea Level Static (Standard Day)			

All fluids are compressible (all fluids ¡¡¡¡); so is possible to considerer the named factor compressibility:

$$K = \frac{\partial p}{\partial V / V} = \rho \left(\frac{dp}{d\rho} \right)$$

If we suppose that the density is constant (is a concept mathematical), we assume that the divergence is zero; give a sphere ratio "a", with "P0" center and "P" point in this sphere ("v" velocity):

$$Force = F = \rho v$$

$$Sa = surface(sphere)$$

"Ba" is sphere; "V" is volume:

$$\iint_{Sa} \vec{F}^d \vec{S} = \iiint_{Ba} div\ (F\ (P))\ dV \approx \iiint_{Ba} div\ (F\ (P0))\ dV$$

$$\iint_{Sa} \vec{F}^d \vec{S} = div\ (F\ (P0))\ \iiint_{Ba} dV$$

$$div(F\ (P0)) = \lim_{a \to 0} \frac{1}{(4/3)\pi a^3} \iint_{Sa} \vec{F}d\vec{S}$$

$$\iint_{Sa} \vec{F}d\vec{S} = Flux$$

If div>0 exit molecules.
If div<0 entry molecules.
If div=0 then density constant.

The use of this method of atmosphere is often used to exchange aerodynamic information without needing to know the density with which the trial has been made. The trial is supposed to have been made at a standard height of 0 meters above the sea level. Later, we will see more things about it.

The pressure is another important parameter in all aerodynamic studies and depends on the already described density.

There are three kinds of pressure:

- Atmospheric pressure.

- Relative pressure.

- Absolute pressure (the addition of the atmospheric and relative pressure).

- Dynamic pressure.

The atmospheric pressure is the pressure exerted by the air on the Earth (the atmospheric pressure at a point coincides numerically with the weight of a static lift of air of straight unit section which spreads from that point to the top limit of the atmosphere). The relative pressure is the pressure which is there, without considering the atmospheric pressure and its variations; the dynamic pressure, is created by the speed particle; the molecules so, have a gap between them more little, so the density and pressure are bigger. May be considerer to dynamic pressure, a part of static pressure....

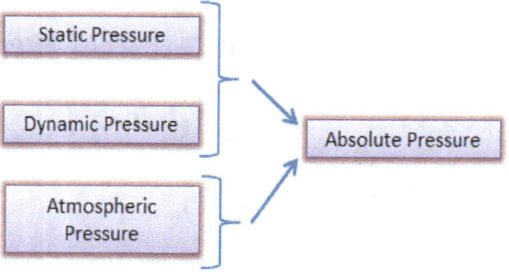

The air density decreases as height increases. That's why you cannot calculate that weight at least you are able to express the variation of the density of the "p" air according to the height "z" or to the pressure "p". Therefore, it is not easy to make an accurate calculation of the atmospheric pressure on a point of earth's surface. On the contrary, it is very difficult to measure it, at least with some accuracy, because both the temperature and the air pressure are continuously varying. We will discuss these interrelated variations in the section of "State equations".

In many studies, the air density is supposed to be constant and, therefore, incompressible (when air pressure increases, the molecules are closer per unit volume and, therefore, the density increases). This is absolutely false: there is no fluid, air or not, that has a constant density. For speeds under 400 km/h (…Mach < 0.3), air density doesn't change a lot (but it changes). All fluids are more or less compressible, but they are.

The fact of assuming that the density is constant, simplifies some calculation processes that, otherwise, would be very laborious, complicated and especially long. Supposing that the density is constant for speeds under 400 km/h, "always" means a mistake, but it is an affordable and small mistake. Furthermore, we can know it. Given a group of fluid particles, we know that each of them can only move within a sphere of a given radius: a particle can be neither further nor closer of a certain distance. Depending on the magnitude of these distances, we will obtain variously compressible fluids.

Everybody knows the game called Newton's cradle:

The ball in motion collides with the row and makes the last ball move indefinitely. The row of balls transmits the movement from the start to the end, from the first to the last ball, making a good example of the conservation of energy and movement quantity. This is due to the rigidity of the balls, but it would be the same if they weren't rigid or solid, as long as the deformations were not permanent and if there weren't hysteresis phenomena. That is, when the balls of the pendulum were perfectly elastic. The sound or vibration is "instantly" transmitted over a piece of metal, due to the compactness of the material. Talking about constant density implies a high power of abstraction. If we put some billiard balls together inside the triangle and try to move them, we won't be able. However, if the balls are tennis balls, we will be able… When we start a game of pool, we hit the white ball towards the first ball. When this ball collides, "all" the balls are ejected. This is due to the impact transmission towards "all the balls".

Similarly, we can think of a stone squeeze: we receive the same force we apply because the stone molecules transmit it. Let's think of a fluid element or a body submerged in a fluid. Let's suppose that we push a portion of fluid. We will obtain the same force in every portion of the fluid. For this reason, the pressure of all the sides of the body is exactly the same. This is why without gravity, a body is in balance when is submerged. Let's analyse and quantify this fact. The distribution of the forces which act on a submerged prism is:

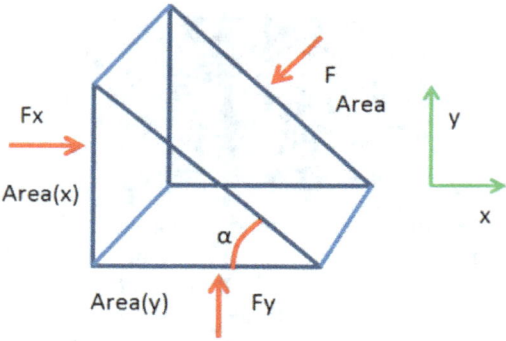

"a" and "b" are the components of a force ("F"), which is applied to the largest surface of the drawn prism. These components are determined by:

$$a = F \cdot \sin(\alpha) \quad b = F \cdot \cos(\alpha)$$

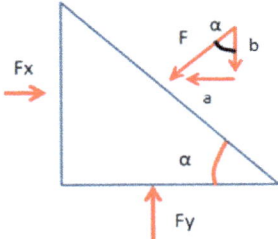

The additions of the forces (components), which act on the prism, are as follows:

$$F_x - F \cdot \sin(\alpha) = 0$$
$$F_y - F \cdot \cos(\alpha) = 0$$

We can write the areas of the prism as follows:

$$Area(x) = Area \cdot \sin(\alpha)$$
$$Area(y) = Area \cdot \cos(\alpha)$$

This helps us to reach the following relation:

$$\frac{F_x}{Area(x)} = \frac{F}{Area} = \frac{F_y}{Area(y)}$$

The pressures are the same on all the faces of the prism. This determines the balance of "internal" forces on any part of the fluid or immersed object.

Another think very important: if the gap space between the molecules is very small, the vibration transmission is faster; "M" Mach number is:

$$M = \frac{V}{c}$$

"c" is the sound velocity in the context.

$$c = \sqrt{\gamma R T} \qquad \gamma = \frac{Cp}{Cv}$$

"R" is de gas constant and "T" the temperature. "Cp" and "Cv" are the specific heat with constant pressure and constant volume.

Also:

$$c = \sqrt{\frac{dP}{d\rho}} = \sqrt{\frac{E}{\rho}}$$

In water for example: $E = 2.15 \cdot 10^9$ N/m^2 with a density = 999.8 Kg/m^3

In water: c=1466.4 m/s sound speed with 0° Celsius.

In solids:

- Concrete: 3200 – 3600 m/s

- Granite: 5950 m/s

- Diamond: 12000 m/s

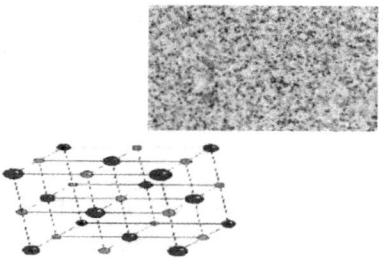

The size or gap between molecules is very important, because define (in CFD) the temporal size or step time; that is:

→ If "c" is the sound speed, mesh-size / step-time = c; so step-time = mesh-size / c; the step time, is normally (necessary) very little.

If the system or problem is adiabatic, is possible to define pressure:

The difference form of the ideal gas law PV=nRT is $P\Delta V + V\Delta P = nR\Delta T$

Since Q=0 for the adiabatic case, the first law of thermodynamics

becomes $\Delta U = -P\Delta V$

and the expression for specific heat $C_V = \dfrac{1}{n}\dfrac{\Delta U}{\Delta T}$ leads to $n\Delta T = \dfrac{-P}{C_V}\Delta V$

Using $R = C_P - C_V$, the gas law gives

$$n\Delta T = \frac{P\Delta V + V\Delta P}{C_P - C_V} = \frac{-P}{C_V}\Delta V$$

With rearrangement this becomes $\dfrac{\Delta P}{P} + \dfrac{C_P}{C_V}\dfrac{\Delta V}{V} = 0$

The limit $\dfrac{dP}{P} + \gamma\dfrac{dV}{V} = 0$ can be integrated to give

$$\ln(P) + \gamma\ln(V) = \text{constant}$$

$$\boxed{\gamma = \frac{C_P}{C_V}}$$

Using log combination rules, this can be rearranged to $\ln(PV^\gamma) = \text{constant}$

So:

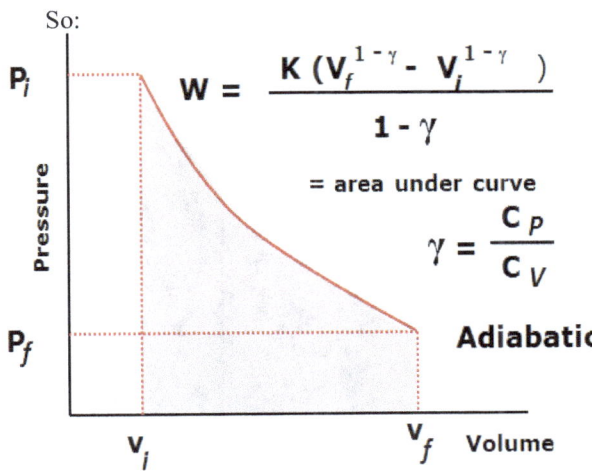

$$W = \frac{K\,(V_f^{\,1-\gamma} - V_i^{\,1-\gamma})}{1-\gamma}$$

= area under curve

$$\gamma = \frac{C_P}{C_V}$$

Adiabatic

From this value, is possible define "P" Pressure (from density and speed sound):

$$P = \frac{\rho C^2}{\gamma} \quad \text{so:}$$

$$\gamma = \log_V \frac{K}{P}$$

$$P = \frac{\rho C^2 \ln(V)}{\ln(\frac{K}{P})}$$

That is: P=f(P) ; is a function of fixed point to resolve.

Now, let's analyze the "buoyancy phenomenon".

It seems a truism but, why does a body float? Archimedes' principle explains that if the weight of the displaced fluid is bigger than the weight of the body, this body will float.
Let's think of an immersed body. This body weighs less (the weight in picture 2 is less than in picture 1):

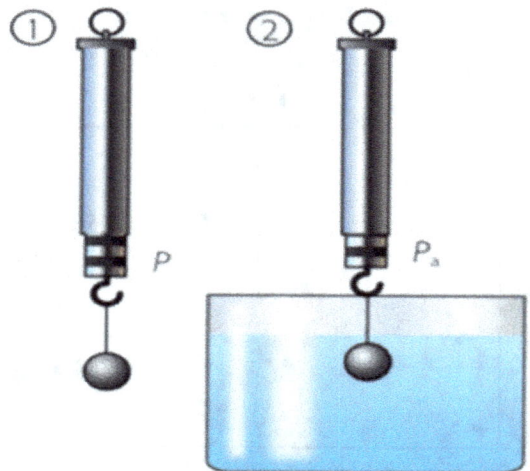

This is because the fluid around or in contact with the ball transmits a force, the vertical component of which is equal to the weight of the dislodged fluid. This fact makes the body weigh less.
Let's think of the same experiment in absence of gravity: neither the ball, nor the water molecules would weigh and, therefore, the ball wouldn't weight.

If we inflated ourselves, we would be able to float in the air:

How much should we inflate ourselves to be able to float in the air?

When asking how much should we inflated ourselves, what we are really asking is how much volume should we occupy, given our weight, to equalize or reduce the density of the fluid in which we are immersed?

Given the density of a person and the air density at sea level:

$$\rho_{person} = \frac{80 \, kg}{0.07 \, m^3}$$

$$\rho_{air} = 1.22 \, kg/m^3$$

The "V" volume that we must reach to float is "x" times the initial volume of a human person, which is around 0.07 m³. If the mass of the person is 80 kg, let's see:

$$\frac{80 \, kg}{0.07 \, m^3 \cdot x} \leq 1.22$$

$$x \geq \frac{80 \, kg}{1.22 \, kg/m^3 \cdot 0.07 \, m^3} \cong 937$$

In other words, we have to become an approximately 66.000 liters ball!

We can see another very illustrative effect of the importance of the density and its existence:

Let's think of a submerged pendulum. We make it swing.

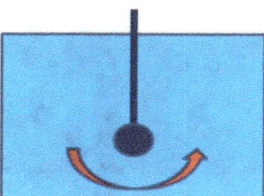

We will be able to see that the pendulum will stop oscillating almost immediately. This is due to the opposition of the water molecules which act on it. In fact, the more density the fluid has (less compressibility), the less time the initial oscillation will take to stop.

Now, let's think of two identical pendulums immersed in a fluid and with opposed oscillations.

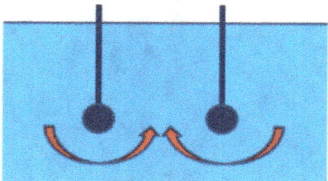

After a short time, both pendulums will oscillate in the same direction and with the same frequency!!!
Why does this fact happen?

Because the density of the fluid, because its variations and the forces transmission trough the molecules. On the moon, this wouldn't happen, due to the air absence.

Once again, we can appreciate the laziness of the nature to the changes...

Some final considerations about the density:

- The higher the density, the higher the *downforce* which is produced by the vehicle.

- The higher the density, the better the motor works (more molecules by unit volume which goes into the engine and, therefore, more useful power).

- The higher the density, the higher the *drag*.

- The lower the temperature, the higher the density will be.

TEMPERATURE

It is a very important variable from an aerodynamic point of view. In fact, the most important thing is that its variation makes, for example, the density and pressure change. From all the temperature measurement units, two of them are the most important:

Kelvin (K) and Celsius degrees (°C); the relation among both measurements is:

$$K = {}^{\circ}C + 273.15K$$

The temperature, according to the kinetic theory of gases, is a function of the average kinetic energy of the molecules which constitute it. In this way, we can say that, as a first approximation (ideal gas), the temperature is determined by the following expression:

$$T = \frac{2}{3} \cdot \frac{1}{K_B} \cdot \langle E_c \rangle$$

"T" is the temperature; "K_B" is the Boltzmann's constant, and Ec is the average kinetic energy.

One of the most important consequences which the temperature has on the air is the fact that the lower the temperature, the most molecules per unit volume there are. This is a damaging fact to the *downforce* (it makes the *downforce* be lower) and to the functioning and performance of the engine.

The movements initials between air molecules, define the full behavior of the air in time; that is; for example, in the Coriolis effect:

A base example of Coriolis effect is the thought experiment in which a projectile fired from the north Ecuador. The canyon is turning to land to the east and therefore the projectile that speed (plus forward speed when the drive).

The projectile traveling north of the earth flies over whose linear velocity eastward decreases with increasing latitude. Inertia projectile east makes its angular velocity increases and thus the points forward to flies. If the flight is long enough, the projectile will land on a meridian east of that from which it was fired, although the direction of the shot exactly north.

Finally, the Coriolis effect, by acting on masses of air (or water) in middle latitudes, induces a twist to divert to the east or west of that mass parties that win or lose altitude latitude in their movement.

Same with the movement of air masses. Cyclones and hurricanes rotate in one direction in the Northern Hemisphere and counterclockwise in the Southern Hemisphere. The reason, again, is that in the northern hemisphere go north means approaching the Earth's axis of rotation and go away to the south, while in the southern hemisphere means go north away from the axis of rotation and go to south approach.

North / south:

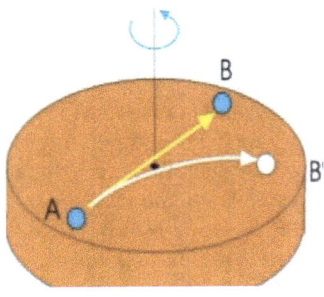

AIR STATE EQUATIONS

All the variables of the air are interrelated. We know, by our experience, that if we climb a mountain or take a plane, the temperature and atmospheric pressure vary. In this section, we will study mathematic expressions and relations among the most significant variables in order to discover its variations depending on other variations. This expressions or relations are called "State Equations".

The most used state thermal equation is the "Clapeyron" equation, due to its simplicity and its good results. It has been obtained through the study of the "Amagat" curves for the ideal gases. If we choose any given gas and represent for different temperatures the function $\dfrac{P \cdot v}{T}$ ("v" is the molar volume and "P" is the pressure), we can observe how, for all the temperatures, the curves meet each other in the same point when the pressure approaches zero. The same happens when we do the representation with different gases. This so interesting cut-off point on the X/Y axis has the value of 8,314 J/mol·K in the international system. Yes, it is the called "R" universal constant of ideal gases.

An ideal gas is a gas whose particles haven't got any potential energy. They only have kinetic energy. In other words, an ideal gas is a gas whose particles don't interact with each other in a perceptible way. This is why the model of ideal gas is used in the cases in which the temperatures are relatively high (the potential energy is insignificant compared to the kinetic energy) or in those in which the density is low (the potential energy of the particles is very low). Fortunately, this happens in most cases. The state thermal equation of an ideal gas is determined by the following expression:

$$P \cdot V = n \cdot R \cdot T$$

"P" is the pressure, "T" is the temperature, "R" is the "universal constant of the ideal gases" (useful for all the gases) and "n" is the number of moles.

Talking about ideal gases, we know that a gas is perfect when its specific heat is constant.

The known Gay-Lussac or Boyle-Mariotte laws are deducible through the state thermal equation of ideal gases, depending on the variables which remain stable in the thermodynamic process.

To simplify the expression and make it particular, we use: $n \cdot R = m \cdot R_g$, where "m" is the mass, and "R_g" is the air constant (it is a constant for our application of low velocities, low pressures and low temperatures).

Therefore, we can write the following expression:

$$P \cdot V = m \cdot R_g \cdot T$$

Important (from this last equation):

If P augments, density and T augment.

If Density is lower, T augment and P is lower.
Etc....

To obtain the variations of each of them, depending on the other variations, we derive the expression, obtaining:

$$d(P \cdot V) = d(m \cdot R \cdot T)$$
$$V \cdot dP + P \cdot dV = m \cdot R \cdot dT + R \cdot T \cdot dm$$

"R_g" (for the air) is: 287.03 J/kg·K. The three differentials belong to the possible variations of such variable.

We can obtain really useful expressions through the state equations. For example, let's think of two points at a different height in comparison to sea level. We can write the following:

$$\frac{P}{P_0} = \frac{\rho}{\rho_0}$$

$$\rho = P \cdot \frac{\rho_0}{P_0}$$

We know, by the basic equation that, being "g" the gravity acceleration:

$$dP = -\rho \cdot g \cdot dy$$

If we substitute, can write the following:

$$dy = \frac{-P_0}{g \cdot \rho_0} \cdot \frac{dP}{P}$$

If we integrate at both sides of the equation:

$$\int_{y_0}^{y} dy' = \frac{-P_0}{g \cdot \rho_0} \cdot \int_{P_0}^{P} \frac{dP'}{P'}$$

$$y - y_0 = \frac{-P_0}{g \cdot \rho_0} \cdot \ln\left(\frac{P}{P_0}\right)$$

We can reach the following expression:

$$P = P_0 \cdot e^{-\dfrac{\frac{y-y_0}{P_0}}{g \cdot \rho_0}}$$

Obtaining the pressure depending on the "y" height, and knowing the pressure at an "y_0" height…

Let's suppose another hypothesis. For example, the relation:

$$T = T_0 + \beta \cdot y$$

We would obtain another state equation for the pressure and height.

This procedure of supposing a hypothesis to make a useful expression is something very usual and necessary in many cases: we have to be able to deduce the correct expression, this expression which allows us to know what we want to know.

For example, let's suppose that we have to know an expression to know the area of a circumference. Firstly, we must decide what values or parameters will participate. In this case, we only choose the radius of the circumference. "A" is the searched area, "R" is the radius and "K" is a constant.

We have to calculate "a":

$$A = K \cdot R^a$$

Now, let's make a dimensional analysis:

$$[A] = [K] \cdot [R^a] \rightarrow L^2 = L^a \rightarrow a = 2$$

Therefore, we have the expression and K=π.

Let's see other example:

Let's suppose that we have to know the density. To do it, we can use one of the two following expressions. We will work with that expression which with we can work easier: (*)

$$\rho = \frac{mass}{volume}$$

From basic equation hydrodynamic (*):

$$\rho = \frac{pressure}{gravity \cdot height} \rightarrow submerged\ body$$

There are a great number of state equations, such as the "Van der Waals" equation, the viral development or the "Berthelot" equation. Nevertheless, the state equation of ideal gases is enough to our action conditions.

For the more skeptical people, we will use the called "Z" compressibility factor, which offers us a measurement of a really good approximation to the ideal gas. This variable is defined as:

$$Z = \frac{P \cdot V}{N \cdot R \cdot T}$$

Therefore, when this variable is equal or very close to the unit, the gas behaves perfectly like an ideal gas. Let's see some values of the "Z" compressibility factor for the air in an exaggerated level of acting:

Temperature (°C)	1 bar	5 bar
-23.15	0.9992	0.9957
26.85	0.9999	0.9987
76.85	1	1.0002

As we can see, the ideal gas model is away from realty when the temperatures go down and the pressures go up. Our chosen range is -23.15 °C and 5 bars, when the compressibility factor is 0.9957. This is not a common situation.
We cannot find this work conditions in any track. As we have seen, even in extreme conditions of temperature and pressure with which we wouldn't work, the ideal gas model is still very useful.

Now, let's see some examples of the use of the "Clapeyron state equation". The following, is a typical case in the racing world:

We have aerodynamic data taken at different times, on the track or in the wind tunnel.

We need to standardize them to 250 km/h in order to compare them well. For example, we can apply it to a GP2 single-seated car category (*downforce* data in kilograms and velocity in km/h).

Config.	Speed (km/h)	Downforce Front (Kg)	Downforce Rear (Kg)	Balance
1	242,84	300,18	550,68	35,22%
2	241,26	301,39	524,63	36,43%
3	241,14	273,18	527,59	34,08%
4	241,17	264,92	511,1	34,07%

The *downforce* is determined by the expression:

$$DF = \frac{1}{2} \cdot \rho \cdot C_z T \cdot A \cdot V^2$$

"ρ" is the air density, "$C_z T$" is the *downforce* coefficient and "A" is the frontal area of the car.

We want to obtain the *downforce front* with a different velocity but with identical conditions.

The quotient between the known *downforce* and that which we want to discover is determined by:

$$\frac{DF_1}{DF_2} = \frac{V_1^2}{V_2^2}$$

$$DF_2 = \left(\frac{V_2}{V_1}\right)^2 \cdot DF_1$$

If we resolve the equation we have:

The subscript "2" indicates the velocity, in this case, 250 km/h.

The subscript "1" indicates the velocity in the table above.

Therefore, to correct, for example, the first velocity, we calculate:

$$DF_2 = \left(\frac{250 \ km/h}{242.84 \ km/h}\right)^2 \cdot 300.18 \ kg$$

As we have said, we have supposed that all the values are obtained with the same density.

In case that we have a different temperature and pressure (different density), we would use the state thermal equation to find the density and do again the aerodynamic charges quotient. In other words:

$$\frac{DF_1}{DF_2} = \frac{\rho_1}{\rho_2} \cdot \frac{V_1^2}{V_2^2}$$

$$DF_2 = \left(\frac{V_2}{V_1}\right)^2 \cdot \frac{\rho_2}{\rho_1} \cdot DF_1$$

If we operate, we have:

As we can see, the procedure is not complicated. Let`s see another typical case of the use of state equations:

We know that the less unsprung mass we have, the better the car dynamics will be. For this reason, we can make the tire air weigh less, keeping a certain pressure.

To do this, we use the studied expressions.

We can create an Excel (for example) sheet in order to do some calculations.

Let's suppose that we want to inflate a tire and reach a certain pressure at a given temperature.

Inflate air		Air	
R		0.2968	Kj/Kg K
STD density		1.2506	Kg/m^3
Initial temperature		25	ºC
Tire pressure		1.4	bar

These input data correspond to dry air. In this way, we can obtain a result of:

- Volume of air to be introduced.

- Mass of air to be introduced into the tire.

We can do it with other gases or even mixtures just changing, "the constant/s of the gas/es". In the case of Nitrogen.

Furthermore, and as a culmination, we can also use the same created template in order to know how much gas we have to remove or introduce in the tire, in the following case:

- Temperature has varied.

- Pressure has varied and we want to keep it or reach another pressure.

This is a typical case when a car stops for a revision of pressures.

We can improve the calculation template as follows:

The parameters with which we have to work are:

- Internal volume of the tire.

- Inflation pressure.

- Temperature of the internal gas.

- Mass of the internal gas.

Calculating any of these values must be possible depending on the other three values; this would be the improvement and the goal of the new worksheet template.

Let's analyze the influence of the temperature variation in a tire.

We have the variations of the pressure (bar) depending on the temperature:

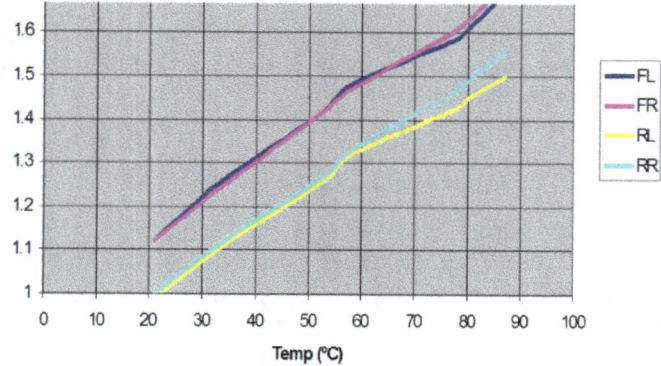

The top two lines correspond to the frontal tires. As we can see, they always tend to heat up more. It is necessary to know these graphs for each tire with which we are working (the damping (bound) or rebound of the tire will change, as well as its diameter).

There is a one expression for air, very important: Sutherland law:

$$\mu = \mu_{ref} \left(\frac{T}{T_{ref}}\right)^{\frac{3}{2}} \frac{T_{ref} + T_s}{T + T_s}$$

$$\begin{cases} T_s = 110.6K \\ T_{ref} = 273.15K \\ \mu_{ref} = 1.716 \cdot 10^{-5} \frac{kg}{m \cdot s} \end{cases}$$

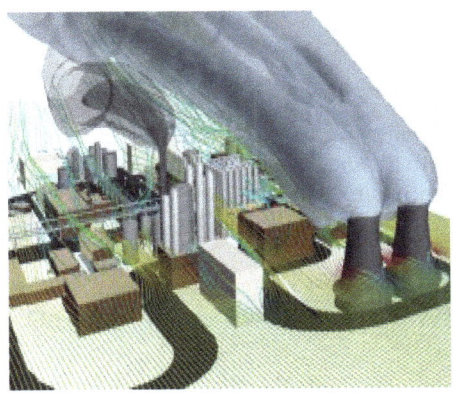

In fact, as an example, Ferrari filling their tires, with a mix between gases:
78% Nitrogen, 21% Oxygen, 1% others, with unknown humidity. Advantages:

Less pressure variation.
Less Work temperature.

Less consume.
Less risc of explosion.
Less oxidation rims.

More: other gas: HFC:

This gas contains 52% of tetrafluoroethane, pentafluoroethane 44% and 4% of trifluoroethane. This mixture is known as GFC 404A.

The HFC gases have the property of driving the rotation by convection heat generated during rotation of the tire to the rim to an almost uniform pressure. The rim acts as a radiator, heat dissipating. It is especially effective in light-alloy wheels used in F1, aluminum and magnesium, preventing overheating of the tire.

The most notable advantages of the use of HFCs are:

Higher durability of the tire.
Temperature almost constant tire.
Possibility of using softer rubber compounds.
Benefits in long term as NASCAR or Le Mans.

→ Note: is possible to obtain one air state equation, from kinetic theory of air: that is:

the measured pressure of a gas arises from collisions of the gas particles with the walls of the container. By considering these collisions more carefully, we can use kinetic theory to relate the pressure directly to the average speed of the gas particles. Firstly, we will determine the momentum transferred to the container walls in a single collision. The figure below shows a particle of mass m and velocity v colliding with a wall of area A. Before the collision, the particle has velocity v_x and momentum mv_x along the x direction. After the collision, the particle has momentum $-mv_x$ along the x direction (note that the components of momentum along y and z remain unchanged). Since momentum must be conserved during the collision, and the momentum of the particle has changed by $2mv_x$, the total momentum imparted to the wall must also be $2mv_x$.

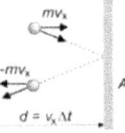

The next step is to determine the total number of collisions with the wall in a given time interval Δt. During this time interval, all particles within a distance $d = v_x \Delta t$ of the wall (and travelling towards it) will collide with the wall. Since the area of the wall is A, this means that all particles within a volume $Av_x \Delta t$ will undergo a collision. We now need to work out how many particles will be within this volume and travelling towards the wall. The number density of the molecules (i.e the number of molecules per unit volume) is

$$\text{number density} = \frac{N}{V} = \frac{nN_A}{V}$$

where N is the number of molecules and n the number of moles in the container of volume V. The number of molecules within our volume of interest, $Av_x \Delta t$, is therefore just the number density multiplied by this volume. i.e.

$$\text{number of molecules} = \frac{nN_A}{V} Av_x \Delta t$$

Since the random velocities of the particles mean that on average half of the molecules in the container will be travelling towards the wall and half away from it, the number of molecules within our volume travelling towards the wall is half of the above value. The total momentum imparted to the wall is now just the momentum change per collision multiplied by the total number of collisions.

$$\text{number of molecules} = \frac{nN_A}{V} Av_x \Delta t$$

Since the random velocities of the particles mean that on average half of the molecules in the container will be travelling towards the wall and half away from it, the number of molecules within our volume travelling towards the wall is half of the above value. The total momentum imparted to the wall is now just the momentum change per collision multiplied by the total number of collisions.

$$\Delta p_x = (2mv_x)\left(\frac{1}{2}\frac{nN_A}{V} Av_x \Delta t\right) = \frac{nMAv_x^2 \Delta t}{V}$$

where we have used $M = mN_A$.

Pressure is defined as the force per unit area, so we need to convert the above momentum into a force in order to calculate the pressure. We can do this using Newton's second law of motion.

$$F_x = ma_x = m\frac{dv_x}{dt} = \frac{dp_x}{dt}$$

Applying this   we obtain

$$F_x = \frac{dp_x}{dt} = \frac{\Delta p_x}{\Delta t} = \frac{nMAv_x^2}{V}$$

The pressure is therefore

$$p = \frac{F_A}{A} = \frac{nMv_x^2}{V}$$

Finally, there is a small amount of 'tidying up' to carry out on this expression. Since we have based our arguments on a particle with a single velocity v_x, and in reality there is a distribution of velocities in the gas, we should replace v_x^2 with $<v_x^2>$, the average of this quantity over the distribution. We can simplify things still further by recognising that the random motion of the particles means that the average speed along the x direction is the same as along y and z. This allows us to define a root mean square speed

$$v_{rms} = [\,<v_x^2> + <v_y^2> + <v_z^2>\,]^{1/2} = [3<v_x^2>]^{1/2}$$

such that $<v_x^2> = \frac{1}{3} v_{rms}^2$

Our final expression for the pressure is therefore

$$p = \frac{1}{3}\frac{nMv_{rms}^2}{V} \qquad \text{or} \qquad pV = \frac{1}{3} nMv_{rms}^2$$

Since the average speed of the molecules is constant at constant temperature, note that by our simple treatment of collisions with a surface, we have in fact just derived Boyle's law.

$$pV = \text{constant} \quad \text{(at constant temperature)}$$

From this point, it is fairly straightforward to go one step further and derive the ideal gas law. Recall that the equipartition theorem states that each translational degree of freedom possessed by a molecule is accompanied by a ½ kT contribution to its internal energy. Each molecule in our sample has three translational degrees of freedom. Also, because in the kinetic model, the only contribution to the internal energy of the system is the kinetic energy ½ mv_{rms}^2 of the molecules, we therefore have:

$$\frac{3}{2} k_B T = \frac{1}{2} mv_{rms}^2$$

Multiplying both sides through by Avogadro's number, N_A, and rearranging slightly gives

$$RT = \frac{1}{3} Mv_{rms}^2$$

Finally, substituting this resul

$$pV = nRT$$

VISCOSITY

Viscosity is one of the most important variables that characterize the air. It is responsible for many of the phenomena and forces that take place in the car world. Its analysis or quantization is quite simple:

Let's suppose we have two plates. They are separated at an "h" distance. It is no air between the plates. Let's suppose that we move one plate with respect to the other with a "U" velocity.

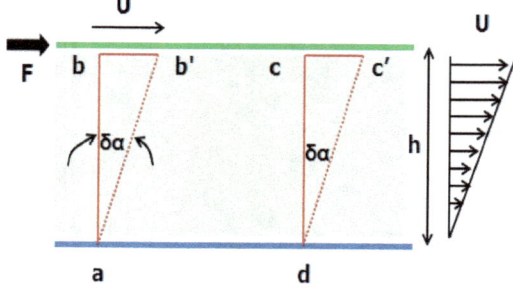

"F" is the force we have to do in order to move the plates. This force is proportional to the "U" velocity and to the "A" area (the area on which we do the force). This force is inversely proportional to the "h" distance which separates the plates. In other words:

$$F = \mu \cdot \frac{A \cdot U}{h}$$

"A" is the area of the plates. We can establish:

$$\tau = \frac{F}{A}$$

$$\frac{du}{dy} = \frac{U}{h}$$

We define the "μ" absolute or dynamic viscosity as the constant of proportionality:

$$\tau = \mu \cdot \frac{du}{dy}$$

We define "kinematic viscosity" as (it is just useful for convenience, gathering two terms):

$$\upsilon = \frac{\mu}{\rho}$$

Viscosity acts as a brake close to the surface, slowing the air molecules. We can see its effect in the following experiment: the air "drags" the cart with a certain force.

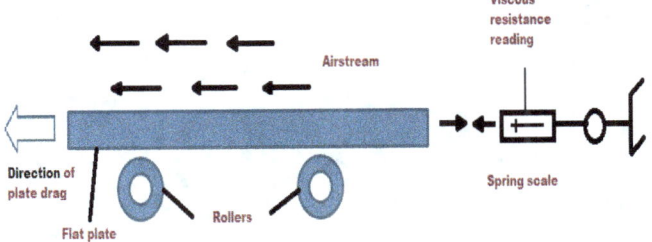

We have to think of viscosity in a different way. Let's see:

We can arrive at the following expression from the definition of velocity and using simple trigonometry:

$$\tau = \mu \cdot \frac{d\alpha}{dt}$$

As we can see, the group of molecules has temporal properties of reaction.

In other words: let's suppose that a certain molecule moves on (for whatever reason) in a certain direction. The molecules around it will react to this change of position, changing its positions and velocities. This will cause a moment exchange, following the first moved molecule. The reaction time of the molecules responding to changes of other molecules is called viscosity:

The longer the reaction time, the higher the viscosity.

We all have said phrases like "traffic in a big city is viscous". This means that cars are "lazy" responding to the changes of the cars which are before them. This brings the slowing of all the cars in the city.

If we manage to think and identify airflow with a "people" flow, we will be able to understand many things.

An essential concept in vehicles aerodynamic is called "viscous plug". It takes place in small cracks or openings such as the suction intake, but especially between the car underfloor and the pavement. Due to the viscosity among the air molecules, there is a moment (which depends on the velocity and air parameters) when a plug is created and the air cannot pass with a high velocity. At that time, the velocity increase is paralyzed despite the fact that the car is faster and faster. We will see this in the section of the diffuser and the ground effect.

Viscosity is a property that varies with temperature. For example, gases viscosity is determined by the following expression:

$$\mu(T) = \mu_0 \cdot \left(\frac{T}{293\ K}\right)^n$$

Between $250\ K \leq T \leq 1000\ K$

As we can see, it depends on the temperature and the viscosity at 20°C (μ_0).

Depending on the kind of gas with which we are working, each parameter will have a different value. We can see the values in the table:

Gas	$\mu_0 \left[\frac{kg \cdot s}{m}\right]$	n	$R[m^2 \cdot s^2/°]$
Dry air	$1.80 \cdot 10^{-5}$	0.67	287
Nitrogen	$1.76 \cdot 10^{-5}$	0.67	297
Oxygen	$2.00 \cdot 10^{-5}$	0.69	260
Steam	$1.02 \cdot 10^{-5}$	1.15	461

In the case that the fluid is liquid, the values of the density and viscosity will be determined by:
This is the expression for the "ρ" density:

$$\rho(T) = a - b \cdot T$$

"a" and "b" are experimentally determined parameters.
The expression for the water is the following:

$$\rho(T) = a - b \cdot |T - 4 \,^{\circ}C|^{1.7}$$

Fluid	μ of $[\frac{kg \cdot s}{m}]$	C	$a\,[\frac{kg}{m^3}]$	$b\,[\frac{kg}{m^3\,^{\circ}c}]$
Water	$1.79 \cdot 10^{-3}$	-	$1.00 \cdot 10^{-3}$	$1.78 \cdot 10^{-2}$
Ethanol	$1.20 \cdot 10^{-3}$	5.72	806	0.85
SAE 10 Oil	$1.04 \cdot 10^{-1}$	15.7	881	0.56
SAE 30 Oil	$2.90 \cdot 10^{-1}$	18.3	902	0.57
SAE 50 Oil	$8.60 \cdot 10^{-1}$	20.2	914	0.59

We show below the values of the parameters which are required for some substances:

$$\mu(T) = \mu_0 \cdot exp\left[C \cdot \left(\frac{293\ ^aC}{273\ ^oC + T} - 1\right)\right]$$

"μ" is the viscosity measured in kg/m/s, "T" is the temperature measured in "°Cᵃ, "μ₀" is the viscosity at 20 °C and "C" is a dimensionless parameter.

The parameters "μ₀" and "C" are experimentally determined for each kind of liquid.

Water is an exception obtained by the following expression:

$$\mu(T) = \mu_0 \cdot \left[7.003 \cdot Z^2(T) - 5.306 \cdot Z(T) - 1.704\right]$$

Where $Z(T) = \dfrac{273\ ^oC}{273\ ^oC + T}$ and $\mu_0 = 1.788 \cdot 10^{-3} \dfrac{kg \cdot s}{m}$

-- If the density is zero, the viscosity is zero: thinking about....

THE ATMOSPHERE

The atmosphere is a gaseous layer which is approximately 10.000 km thick and is located around the Earth. It's composed by different gases –principally nitrogen and oxygen- and solid and liquid particles in suspension which are attracted by the Earth's gravity.

All the weather and meteorological phenomena are produced in this layer. It regulates the energy input and output from the Earth.

It's the main way of heat transfer. The atmospheric pressure value is the same as the weight of a unitary air column from the point where we want to calculate the atmospheric pressure to the atmosphere upper limit.

Due to the weight of the particles which form the atmosphere, the highest percentage of the atmospheric mass is located in the first kilometers. 50% of it is located under the 5 km, 66% is located under the 10 km and beyond the 60 km only one thousandth of it is located.

Therefore, as we can see, it is not a gaseous layer with constant density, but it is a layer whose density decreases when the height from ground increases.

Taking a linear fall of the temperature in the atmosphere such $T = T_0 - L \cdot h$ is a good approximation for the first 11 km of the atmosphere, called Troposphere.

Meanwhile, in the first 9 km of the Stratosphere, the temperature remains constant at -56.5 ° C, as we can see in the following figure:

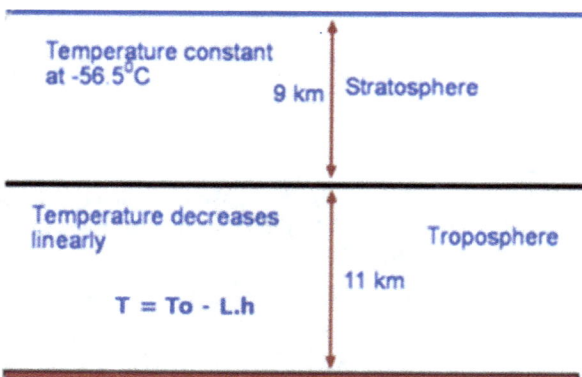

The temperature variation depending on the height for the whole atmosphere is shown in the following graphic:

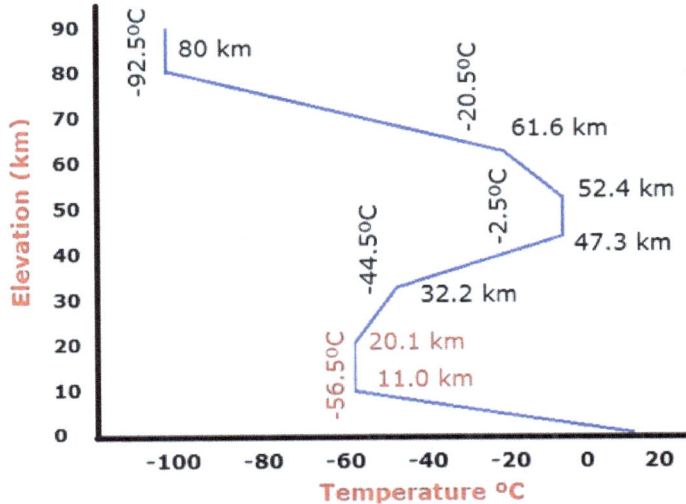

All the values we have seen depend on each other. This makes it necessary to determine a common and standard atmosphere to be able of unify and contrast experiments, tests, trials and calculations. This base atmosphere is called "Standard Atmosphere".

Surface values at sea level:

- Temperature: 15ºC (59ºF).
- Pressure: 760 mm o 29,92" of mercury column, equal to 1013,25 mb per cm².
- Density: 1,225 kg/ m³.
- Acceleration due to the gravity: 9,8 m/s².
- Speed of sound: 340,29 m/s.
- Thermal gradient of 1,98 ºC per each 1000 feet or 6,5 ºC per each 1000 m.
- A pressure drop of 1" per each 1000 feet, or 1 mb per each 9 meters, or 110 mb per each 1000m.

LAMINAR FLOW AND TURBULENT FLOW

All of us know what laminar and turbulent mean, but hardly anyone knows its mathematical definition.

"Laminar" sounds like "light" or "linear". "Turbulent" sounds like "complex". It is indeed thus. We say that the flow moves in a laminar way or it is somewhere, laminar, if the velocity field rotational is zero. If it isn't, we say that it is a turbulent flow. In other words:

$$rotational \; \bar{v} = \nabla \; x \; \bar{v} = \begin{vmatrix} i & j & k \\ \dfrac{\partial}{\partial x} & \dfrac{\partial}{\partial y} & \dfrac{\partial}{\partial z} \\ v_x & v_y & v_z \end{vmatrix}$$

$$\begin{cases} \nabla \; x \; \bar{v} = 0 \; laminar \; flow \\ \nabla \; x \; \bar{v} \neq 0 \; turbulent \; flow \end{cases}$$

The beginning of the turbulent layer or turbulent flow is something very important that determines the air aerodynamics. It is not easy to know this transition.

We know that the higher the velocity, the more turbulent the flow, but it is complicated to know at what velocity the flow will be turbulent. The Reynolds number can indicate under some certain specific conditions, this step from laminar to turbulent flow.

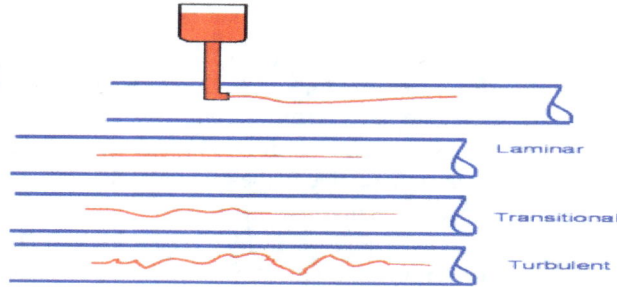

Turbulent flow – velocity profile:

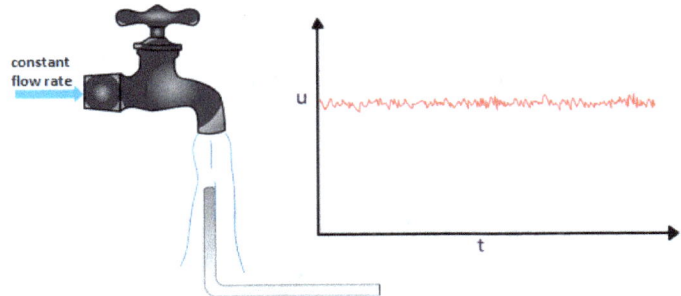

The Reynolds number is defined as:

$$Re = \frac{Inertial\ force}{Viscous\ force} = \frac{\rho \cdot L \cdot v}{\mu}$$

In the case of the cars, knowing the transition velocity is much more complicated. It is necessary to test by using CFD, wind tunnels or similar ones.

For example, for a blue whale in ocean, the $RE=3*10^8$, man swimming $RE=4*10^6$, etc....

There are two explanations of the conversion from laminar to turbulent flow:

Shearing layer (stress layer):

A shearing layer is an area of flow where the velocity gradients are high. In other words, the velocity varies in a very perceptible way, as we advance forward the normal direction or in a perpendicular way to the movement.
Within this flow, an infinitesimal perturbation is caused. This perturbation causes a light wave.
This undulation causes:

a) The increase of the flow velocity on the convex zones (A, B', C, D').

b) The decrease of the velocity on the concave zones (A', B, C', D).

c) If we consider the stable flow, by applying the "Quantity of Movement Equation", it is created a force, which increases the disturbances. In this way, the shearing layer becomes unstable and the original undulations become vortex or whirlwinds.

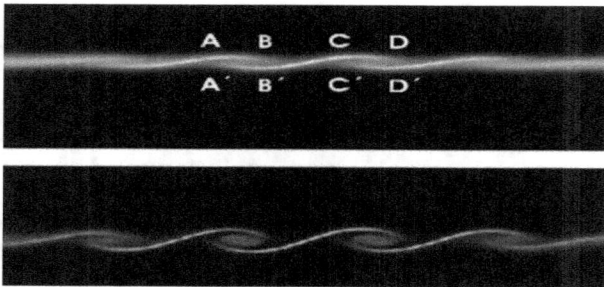

In the real life, if we try to know an event laminar, that will be impossible; always will be a person witch think in turbulent mode; this person, in time, do that all system work in turbulent...

In the real life, we can see these turbulences watching the sky:

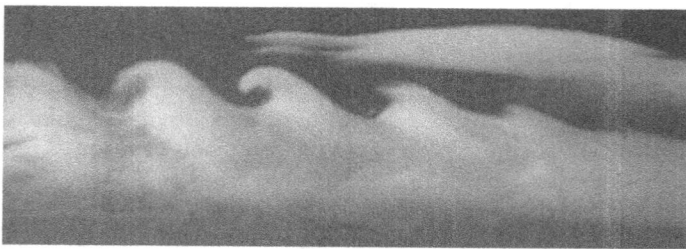

Considering the origin of turbulence in terms of small initial disturbances, one case where we can see and observe the creation of turbulences is the curtains of most rural houses. We all have seen this curtains which are placed on the door to prevent the entry of mosquitoes. If it's windy, we will see that the curtain starts to ripple. Originally, the curtain doesn't move, but with a slight alteration, the wave starts.

In fact, this conception of the origin of the turbulence can create Jupiter's cloud bands and its storms or whirlwinds.

Later, we will study it in depth, but this is the main reason of the necessary precision in the limit layer definition, when we perform a CFD simulation!

Viscosity:

Let's imagine a pedestrian who is being chased by another pedestrian. We suppose that the pursuer pedestrian has a certain reaction time in face of the changes of direction of the chased pedestrian. If that reaction time depends on limits or intervals, the pursued pedestrian's trajectory may trace the following graphic with turbulence aspect:

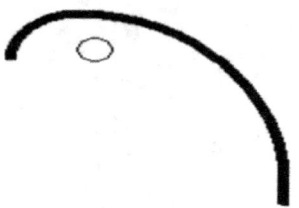

Therefore, depending on this reaction time, the turbulence will exist or not.

Characteristics of turbulent flows:

- **Irregularity:** Any turbulence has a pattern of irregularity because at least apparently, is unpredictable. The fluctuations appearance of fluid-dynamic variables (velocity, pressure, temperature, concentration) with very different sizes and times (different scales), give the turbulence an irregular nature.

- **Three-dimensionality**: Any turbulence is three-dimensional. The lower the scale size, the more perceptible this fact will be.

- **Diffusivity:** The transport phenomena of mass, movement quantity and energy, are increased by the turbulence effect. The corresponding effect is analogous to molecular scales with the molecular transport.

- **Dissipation:** After the turbulent flow has been developed, the turbulence tends to be stable, being necessary an input of energy. This energy becomes a series of processes of fluid elements deformation. The absence of this energy input means that the turbulence is progressively smaller.

Sir Horace Lamb (1849-1934), in an international tribute which was celebrated for his eightieth birthday in 1929, said: "When I die, I hope I go to heaven. There, I hope being illuminated about the solution of two problems: quantum electrodynamics and turbulence. I'm very optimist about the first one...".

The first problem was solved by Richard P. Feynman (1918-1988) and because of that, he won the Nobel Prize in 1965. Feynman: "Turbulence is the last important unsolved problem of classical physics".

BOUNDARY LAYER

DEFINITIONS

The concept of "boundary layer" defined by Prandtl (1905) necessarily implies two conditions: the first is the non-slip condition: the fluid in direct contact with the surface must be at rest.

The second is that the shear forces between layers of fluid must be non-zero (skin friction drag).

The boundary layer is a direct consequence of the viscosity of air: if there were no viscosity, there wouldn't be a boundary layer and therefore no pressure distribution on an object, neither the downforce as we know it.

Therefore, it is a very very important concept.

As we have seen, the velocity of the air molecules which are close to a surface is increasingly lower as we approach the surface.

If we move away from the surface, the fluid elements increase its velocity up to a distance where the air velocity is not affected by the vehicle.

When the molecules are at a distance where its velocity is 99% of the relative velocity between surface-air, we define this distance as "Boundary layer" and its "δ" thickness as the boundary layer thickness in that point.

There are other definitions but these are the most used.

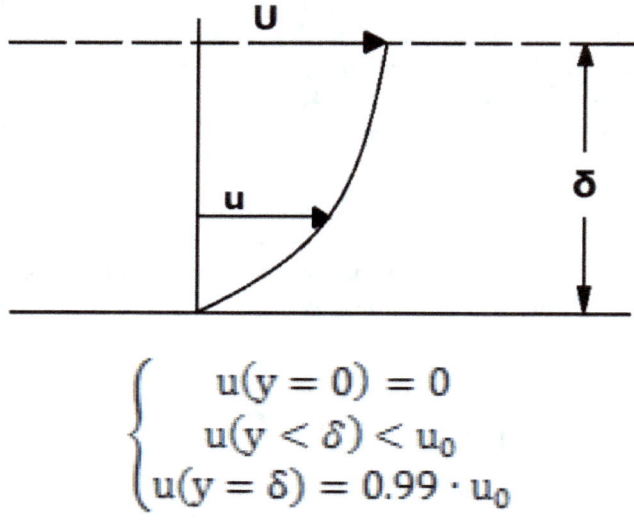

$$\begin{cases} u(y = 0) = 0 \\ u(y < \delta) < u_0 \\ u(y = \delta) = 0.99 \cdot u_0 \end{cases}$$

This can be mathematically written in the following way:

Where "u(y)" is the fluid velocity which depends on the distance from the surface, "u_0" is the relative surface-flow velocity in areas which are sufficiently distant so that the fluid velocity is not affected by the vehicle.

Pilots who fly a paraglide know that when the winds are strong, the rule is to fly at a low altitude. Why? Because the closer to the ground you are, the lower the velocity of the wind. This is due to the boundary layer:

As we have said, the fluid velocity on the surface is zero. We can see this happening every day on the highways. ¿Why isn't dust on cars removed while driving them? The dust is not removed by the air because the air layer in contact with the vehicle surface (this layer which would tend to push the dust particles) has zero velocity. Therefore, we should not wait for the wind to clean our vehicle:

Besides this definition of boundary layer thickness, based on the condition: $u(y=\delta)=0.99 \cdot u_0$, we have other definitions where the percentages of the "u_0" velocity are different.

Regarding the velocity profile, Blasius proposed that the velocity of the different fluid layers on the boundary layer (if this layer is laminar), is independent of the distance to the other end of the layer and it is function of the proportion y/δ. The same happens with the following relation:

$$\frac{u(y)}{u_0} = sin\left(\frac{\pi}{2} \cdot \frac{y}{\delta}\right)$$

We could approximate this expression with a Taylor series, when necessary.

$$\frac{u(y)}{u_0} \sim \left(\frac{\pi}{2} \cdot \frac{y}{\delta}\right) - \frac{1}{3!} \cdot \left(\frac{\pi}{2} \cdot \frac{y}{\delta}\right)^3 + \dots$$

We have taken only the first two terms because it is always true that:

Because of this:

$$\left(\frac{\pi}{2}\cdot\frac{y}{\delta}\right) \leq \frac{\pi}{2}$$

$$\frac{y}{\delta} \leq 1 \quad 0 \leq y \leq \delta$$

Finally, we approximate the profile using a parabolic function:

$$\frac{u(y)}{u_0} = 2\cdot\frac{y}{\delta} - \left(\frac{y}{\delta}\right)^2$$

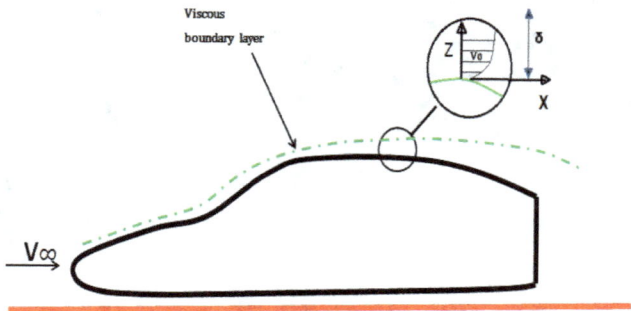

Actually, the boundary layer thickness depends on the distance to the trailing edge of the plate. We can approximate it looking at the dependence with $x^{1/2}$ when this is laminar. In other words: $\alpha x^{1/2}$. The expression is written below:

$$\delta = 5\cdot\left(\frac{u_0}{\nu}\right)^{-1/2}\cdot x^{1/2}$$

The thickness "δ" of the boundary layer:

- Decreases when the viscosity decreases.

- Decreases with the increase of the Reynolds number. Therefore, it also decreases with the velocity.

To show additional pictures of the boundary layer, a CFD simulation of a flat plate has been performed. Some advantages of this method are for example, the capability of showing contours of the velocity around the plate and even velocity vectors of the boundary layer.

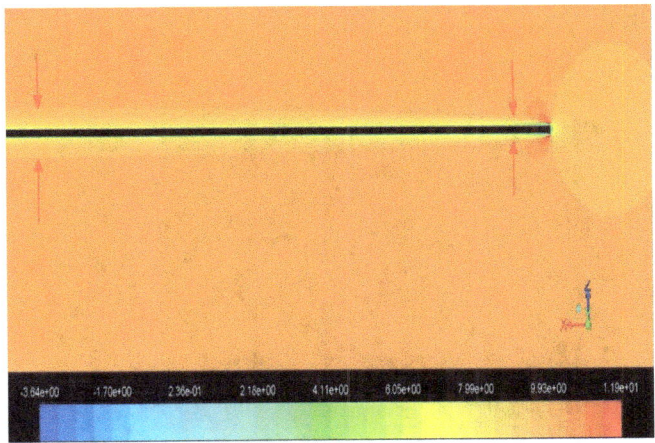

The first picture shows the velocity contours which can be useful to see how the boundary becomes thicker. In this case the air flows at 10 m/s from right to left, it has positive velocity in the x axis. As we know, in the boundary layer the velocity varies from 0, to the free stream velocity, 10 m/s in this case, corresponding to the predominant orange areas.

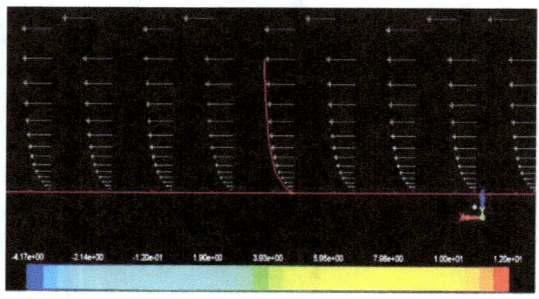

BOUNDARY LAYER

The picture shows how the boundary layer thickness increases as the x component of the plate increases, being thicker in the left part of the image.

In the second picture we can see the velocity profiles close to the plate, which are the same as the ones explained before for the boundary layer.

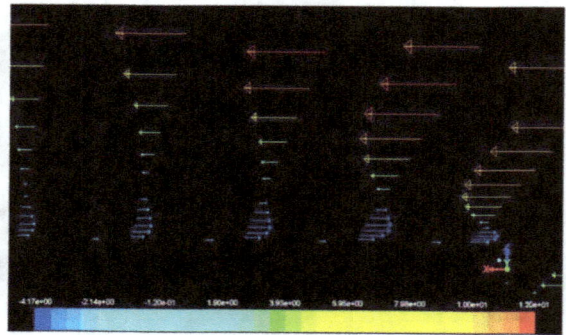

The last example corresponds to a velocity profile affected by an adverse pressure gradient, some vectors show that the flow goes from left to right, contrary to the free stream direction being set from right to left. The air 'sees' a higher pressure zone in the left which makes it move the other direction. Mesh for this simulation:

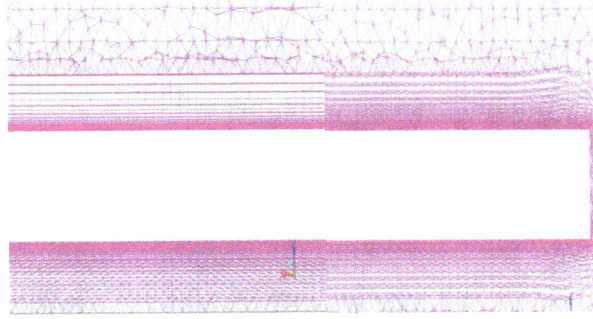

We can see the effect of speed reduction in a wall, trough CFD simulation:

Wall without MOVING WALL in up-down wall:

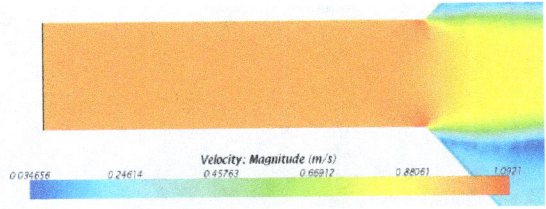

Wall with MOVING WALL in up-down wall:

As we travel along the surface, the boundary layer is thicker and thicker. Within this layer, the most important phenomena (from the point of view of the lift and drag creation) take place.

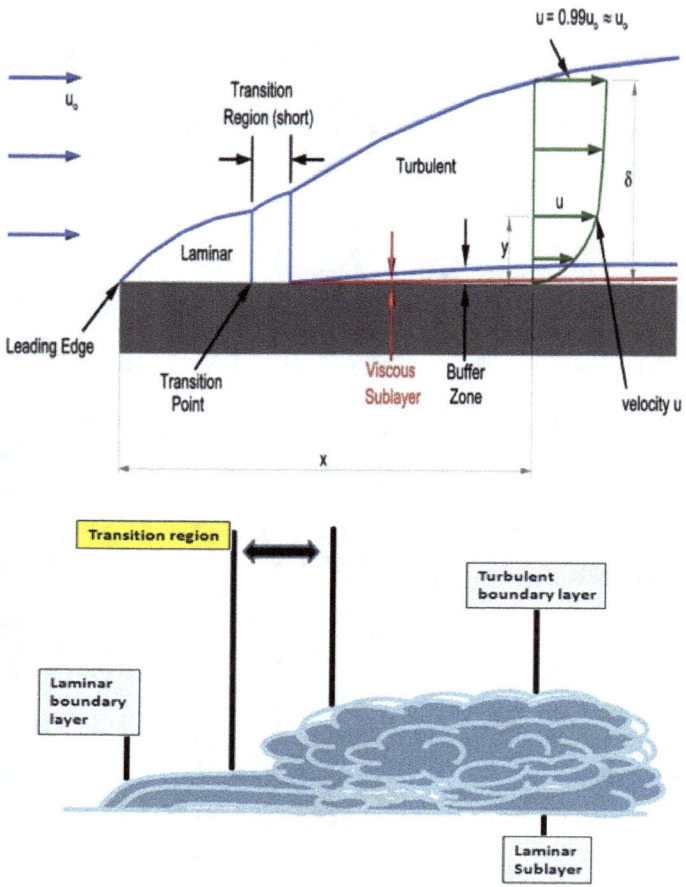

If the boundary layer "gets out of control", lift decreases and drag increases.

Due to this reason, the control of the boundary layer is essential. Keeping the boundary layer close to the surface despite the fact that the surface moves and acquires different configurations, is something important to create lift or *downforce*.

In other chapters, we will see and analyze many ways of keeping this layer attached to the surface or, "on purpose", separate it...

This velocity variation with respect to the distance to the surface can be clearly appreciated in the water circulation through a pipe: as the flow advances, the boundary layer is wider and wider till both limits (upper and lower) meet each other:

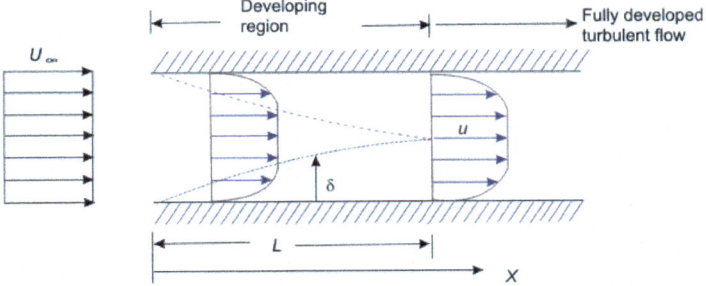

This is something important that we must not forget:

The "useful" fluid pipe section is reduced. It looks as if the pipe's diameter is being reduced. It is something that we mustn't forget, because it determines the sizing of inlet and cooling ducts, etc... We will see it in other chapters of this book.

The transition between the laminar boundary layer and the turbulent layer takes place when the fluid starts to become "unstable", creating sporadic vortex. This point location depends on the Reynolds number, on the density, on the velocity "u_0" and on the viscosity:

$$x_{transition} = \frac{Re_x \cdot \mu_0}{\rho_0 \cdot u_0}$$

This transition is generated at different velocities. In the case of a car, the transition comes early when the velocity is higher. ("TP" is the point of transition, "L" is laminar flow and "T" is the turbulent flow).

We can see it in following image for low speed:

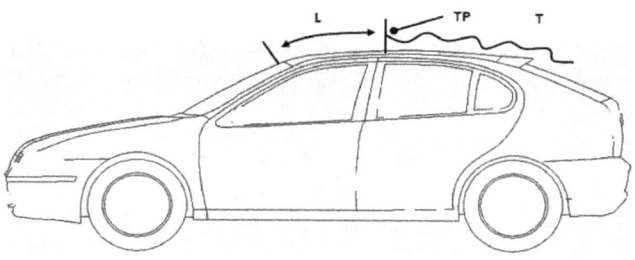

The following drawing is for high speed:

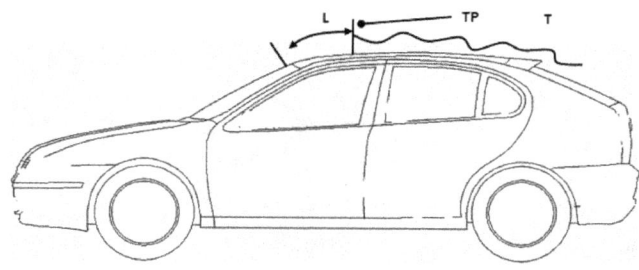

Regarding the Reynolds number, for a semi-infinite plate:

$$Re < 6 \cdot 10^4, \quad \text{Laminar boundary layer}$$
$$Re > 4 \cdot 10^6, \quad \text{Turbulent layer}$$

Therefore, the Reynolds number for laminar-turbulent transition varies approximately between $6 \cdot 10^4$ and $4 \cdot 10^6$ in a semi-infinite plate.

For surfaces with a strong geometric gradient, (in other words, a high variation in curvature), the boundary layer detaches because there is a big shear force. It is said that there is an "adverse pressure gradient". It seems like the molecules which are in contact with the surface pull the other molecules above them, breaking them:

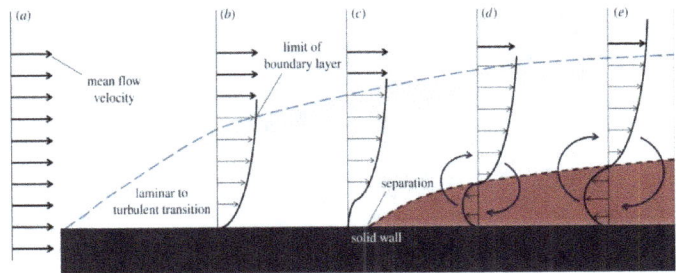

We can see many transition points on a car, depending on what is the shape of the bodywork. In the following images, we can see that the flow of the second car is more uniform and remains close to the bodywork:

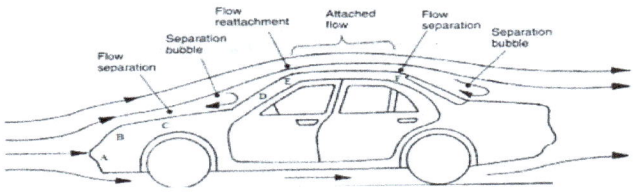

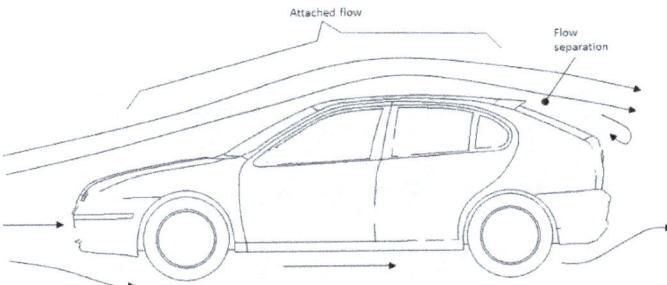

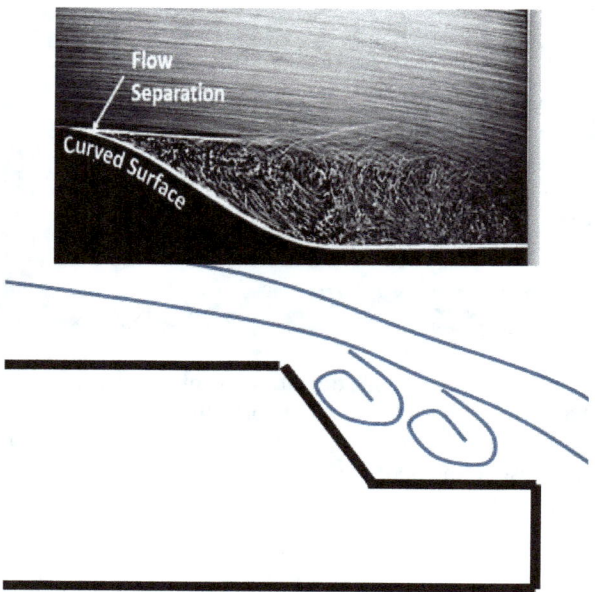

That is (there is not flow separation):

The critical Reynolds number is the value of the Reynolds number in the point in which the boundary layer becomes completely turbulent. Its value for a semi-infinite flat plate (it depends on the surface geometry) is about $4 \cdot 10^6$.

The thickness of the laminar and turbulent boundary layer is determined by the following equations:

$$\delta = \frac{4.91}{2\sqrt{Re_x}} \qquad \delta = \frac{0.382}{5\sqrt{Re_x}}$$

The velocity profile in the section of the turbulent boundary layer is determined by:

$$\frac{u}{u_0} = \left(\frac{y}{\delta}\right)^{1/7}$$

If we analyze the structure of the turbulent boundary layer, we can see two sections. The viscosity effects are strong in the layer which is closer to the surface.

The other layer is more distant to the surface and the turbulence effects exceed in it. The first region (the one dominated by the effects of the viscosity) can be subdivided into two sublayers. The closest to the surface is called linear sublayer and the viscosity effects predominate in it. In the lower sublayer, both effects (viscosity and turbulence) are equally important.

The velocity profile of the linear sublayer (the closest to the surface) is proportional to the distance "y":

$$\frac{u}{u_f} = \frac{u_f}{v} \cdot y$$

Where v is the kinematic viscosity and u_f is the friction velocity, defined by the following expression:

$$u_f = \sqrt{\frac{\tau_w}{\rho}}$$

In this sublayer, as the velocity profile is linear, the velocities gradient is constant:

$$\frac{du}{dy} = \frac{\tau_w}{v}$$

From this expression, we can arrive easily to the other studied expression:

$$\tau_w = \mu \cdot \frac{du}{dy}$$

In the sublayers where the turbulence phenomena are as important as or even more important than the viscosity phenomena, we have to look for another velocity profile. The velocity profile of the more distant layers is determined by a logarithmic law which is expressed as follows:

$$\frac{u}{u_f} = 5. + 5.75 \cdot log_{10}\left(\frac{u_f}{v} \cdot y\right)$$

We know already, that the boundary layer exist, because exist viscosity; that is; if exist friction force between air molecules; see one example about.

Get a section of duct; if the viscosity is no zero, the velocity field is (speed = 10^{-6} m/s):

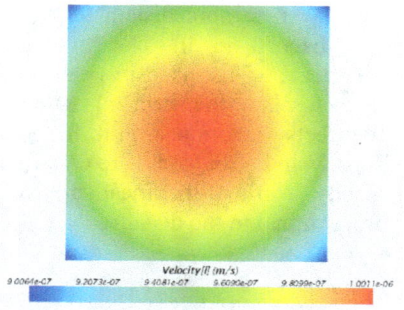

If the viscosity is lower:

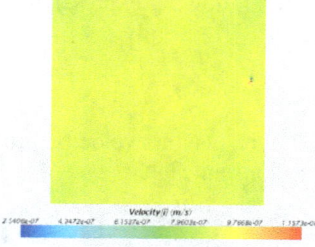

That is so: The fluid with lower viscosity, less braking.

In fact, in this example, the mesh points, is a molecules distribution.... That is: if we applied boundary layer in this simulation CFD, is wrong, because we test the best mesh possible (molecules ¡¡¡¡).

Now let's take a look at the flow detachment on a curved surface.

FLOW DETACHMENT

Adding to what we have already said, the study and understanding of the flow detachment process is one of the most important aspects of aerodynamics. It is enough to say that the flow detachment is the principal cause of *drag* in the "blunt" bodies, such as a plate perpendicular to the flow, a sphere or a vehicle. In this objects, *drag* (due to this phenomenon) dominates the *drag* due to the fluid friction with the surface. The detachment of the boundary layer depends on three aspects:

- The shape of the submerged object.

- The type of boundary layer (laminar or turbulent) that is being generated.

- The pressure gradient; related to the geometric gradient.

The flow detachment is caused by the deficiency of kinetic energy on the boundary layer.

This is due to the loss caused by the viscosity. If we think of a sphere, the flow (after overtaking the upper point of the sphere) tends to loose pressure and gain speed. However, it cannot regain all its velocity, because a loss of kinetic energy (in the form of heat) has been generated due to the viscosity. This causes two things on the velocity profile: first, an inflection point and then, an inversion in the movement of the boundary layer, which causes its separation.

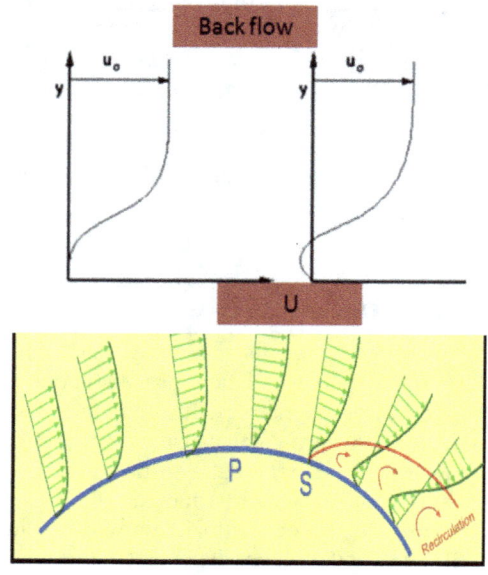

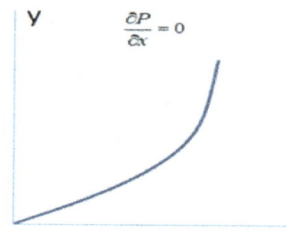

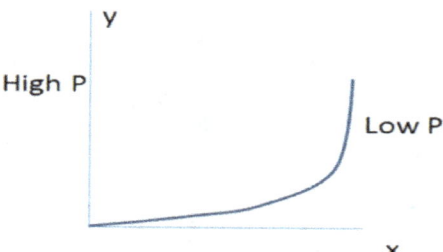

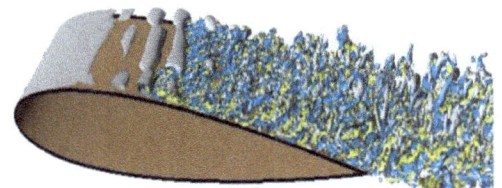

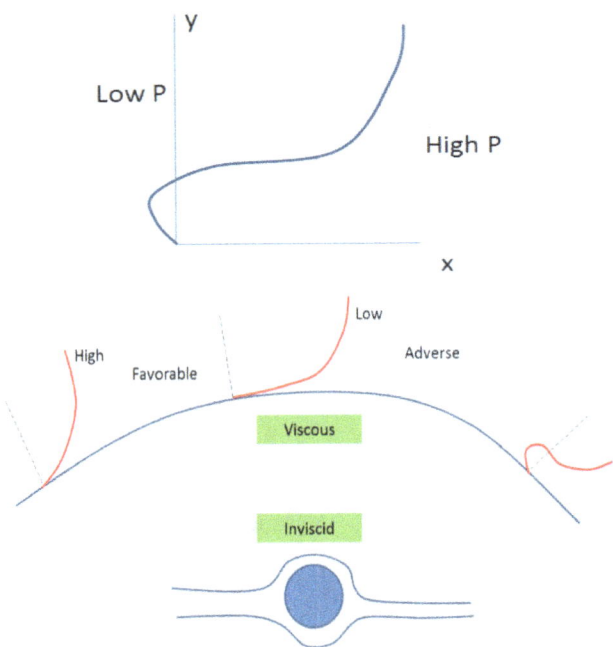

Example of a way for not detaching flow:

We create a suction effect in the trailing edge because, due to Bernoulli effect, the velocity increases in the decreasing area between the profile and the surface, and the pressure decreases. This allows the detachment of the flow over the profile.

At large Reynolds numbers, boundary layers, thin layers of fluid in which viscosity effects are significant, are formed along solid boundaries, because viscous fluids cannot slip at solid boundaries. In the absence of pressure gradients the boundary layer along a flat surface increases in thickness as $l\sqrt{\dfrac{1}{Re}}$. Negative (or favorable) pressure gradients in the flow direction, which accelerate the flow, decrease the boundary-layer thickness and increase the velocity gradient at the wall. Positive or unfavorable pressure gradients tend to decelerate the flow, to increase boundary-layer thickness, and to decrease the velocity gradient at the wall. Unfavorable pressure gradients can cause boundary-layer separation, which often results in drastically altered flow patterns and losses in performance of such devices as airplane wings and diffusers.

At relatively low values of Reynolds number, boundary layers tend to be laminar. At higher Reynolds numbers, a boundary layer is unstable to small disturbances. The disturbances grow, resulting in transition to a turbulent boundary layer. Most practical flow situations involve high Reynolds numbers and turbulent boundary layers. Because of three-dimensional interchanges of momentum, a turbulent boundary layer is thicker and has a larger wall velocity gradient than a laminar layer at the same Reynolds number. The increased momentum near the wall allows a turbulent boundary layer to withstand a larger unfavorable pressure gradient than a laminar layer without separating, but results in higher wall shear stress and drag.

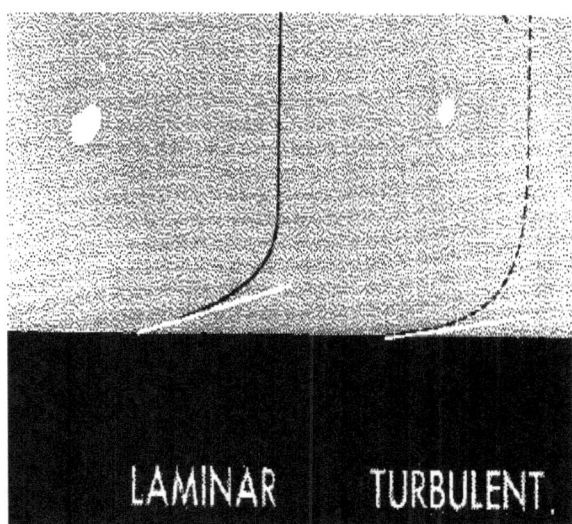

LAMINAR   TURBULENT.

DISPLACEMENT THICKNESS

Let's suppose two situations. In the first situation, we have a real fluid and in the second one, we have a non-viscous fluid. The displacement thickness is the thickness of the fluid layer. This thickness must have null velocity in the non-viscous fluid so that the flow is the same as in the real situation (in which the fluid is viscous), in a way that the conservation law is true for both situations.

Let's consider now incompressible fluids:

The mass flow for a real fluid till a height "Y" is determined by:

$$\dot{m}_{REAL} = \rho_0 \cdot u_0 \cdot (y - \delta) \cdot b + \rho_0 \cdot b \cdot \int_0^\delta u(y) \cdot dy$$

"b" is the thickness of the flat plate. As we can see, there are two contributions for the real fluid: the first element belongs to the mass flow provided by the free circulation section:

$$\rho_0 \cdot u_0 \cdot (y - \delta) \cdot b$$

The second element belongs to the contribution the boundary layer fluid makes to the flow.

$$\rho_0 \cdot b \cdot \int_0^{\delta} u(y) \cdot dy$$

The mass flow for a non-viscous fluid is determined by:

$$\dot{m}_{NON\ VISCOUS} = \rho_0 \cdot u_0 \cdot (y - \delta^*) \cdot b$$

There is only one contribution to the mass flow in the section where the flow moves freely. In the remaining section, the velocity is zero.

U (y)

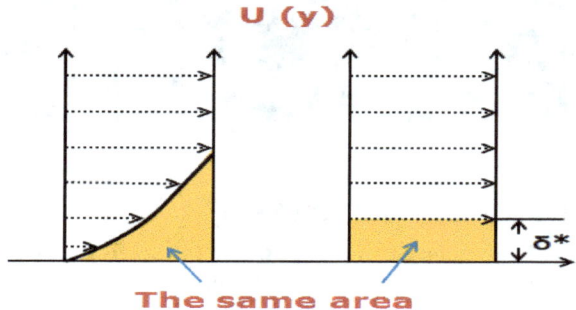

The same area

As we can see in the real fluid, within the boundary layer there are velocities from u=0 to u=0,99·u_0, determined by a certain function y/δ. However, if we want to calculate the displacement thickness in a non-viscous fluid, we have to consider a layer where the velocity is zero until we arrive to δ^*, and then jumps directly to u=u_0.

Due to the term definition, this condition has to be true:

$$\dot{m}_{REAL} = \dot{m}_{NON\ VISCOUS}$$

$$\delta^* = \int_0^{\delta} \left(1 - \frac{u(y)}{u_0}\right) \cdot dy$$

This expression gives us the thickness of the displacement layer.

Let's look at an example. We want to calculate the displacement thickness for the following velocity profile:

$$\frac{u}{u_0} = sin\left(\frac{\pi}{2} \cdot \frac{y}{\delta}\right)$$

The integral would be as follows:

$$\delta^* = \int_0^\delta \left(1 - sin\left(\frac{\pi}{2} \cdot \frac{y}{\delta}\right)\right) \cdot dy$$

Solving the integral:

$$\delta^* = \delta - \frac{2 \cdot \delta}{\pi}\left[- cos\left(\frac{\pi}{2} \cdot \frac{y}{\delta}\right)\right]_0^{\frac{\pi}{2}} = \delta - \frac{2 \cdot \delta}{\pi}$$

Its result:

$$\delta^* = \left[\frac{\pi - 2}{\pi}\right] \cdot \delta$$

We can do the same for different velocity profiles. The parabolic velocity profile is the following:

$$\frac{u(y)}{u_0} = 2 \cdot \frac{y}{\delta} - \left(\frac{y}{\delta}\right)^2$$

The integral would be as follows:

$$\delta^* = \int_0^\delta \left(1 - 2 \cdot \frac{y}{\delta} - \left(\frac{y}{\delta}\right)^2\right) \cdot dy$$

Solving the integral:

$$\delta^* = \delta \cdot \left[\frac{y}{\delta} - \left(\frac{y}{\delta}\right)^2 + \frac{1}{3} \cdot \left(\frac{y}{\delta}\right)^3\right]_0^\delta$$

Its result:

$$\delta^* = \frac{1}{3} \cdot \delta$$

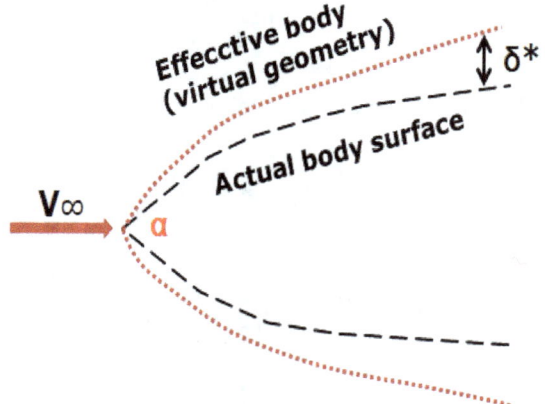

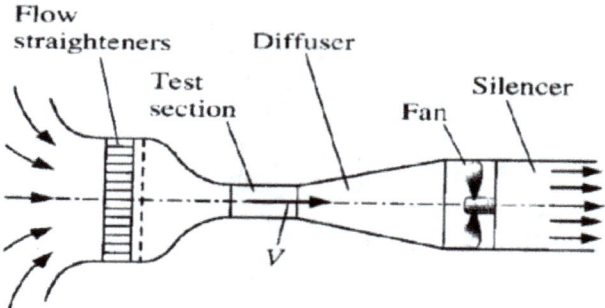

Let's also suppose that we work at 19° C. In this way, we know that the Reynolds number in the testing chamber is:

$$Re_x = \frac{Vx}{\nu} = \frac{4 \cdot 0.3}{1.507 \cdot 10^{-5}} = 7.96 \cdot 10^4$$

This value is lower than the critical Reynolds value (turbulent transition). Its value is $5 \cdot 10^5.$

The displacement thickness can be calculated as follows:

$$\delta^* = \frac{1.72x}{\sqrt{Re_x}} = \frac{1.72 \cdot 0.3}{\sqrt{7.96 \cdot 10^4}} = 1.83 \; mm$$

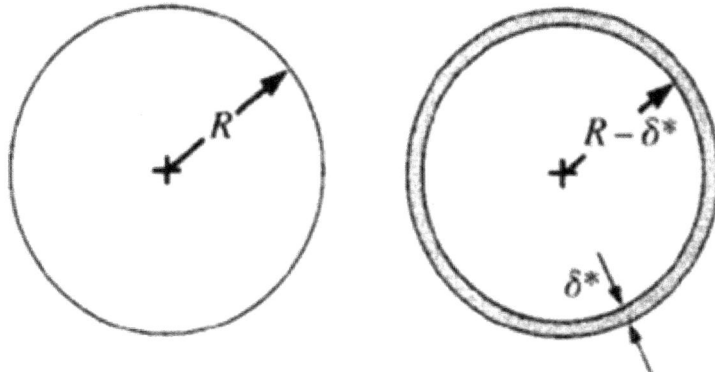

Therefore, the effective radius of the tunnel is reduced in this displacement value. Therefore, the velocity at the exit of the testing chamber is:

$$V_{end} = V_{start} \frac{\pi R^2}{\pi (R - \delta^*)^2}$$

$$V_{end} = 4 \cdot \frac{0.15^2}{(0.15 - 1.83 \cdot 10^{-3})^2} = 4.1 \, m/s$$

The rear average velocity is increased in a 2.5%.

IMPULSE THICKNESS

The impulse thickness is the fluid layer thickness that has null velocity in the non-viscous fluid so that the total quantity of impulse is the same as in the real situation (where the fluid is viscous).

Let's measure, what the impulse deficit, that causes the real fluid viscosity, is in order to find the impulse thickness:

The impulse deficit in the boundary layer of the real fluid is determined by:

$$Impulse \; deficit_{REAL \; FLUID} = u_0 \cdot \int_0^{\delta} \rho(y) \cdot u(y) \cdot \left(1 - \frac{u(y)}{u_0}\right) \cdot dy$$

For a "free" non-viscous fluid:

$$\rho_0 \cdot u_0^2 \cdot \theta$$

As the impulse conservation has to be true in both cases (term definition), we equal the terms:

$$\rho_0 \cdot u_0^2 \cdot \theta = u_0 \cdot \int_0^\delta \rho(y) \cdot u(y) \cdot \left(1 - \frac{u(y)}{u_0}\right) \cdot dy$$

Therefore, the impulse thickness in the real fluid due to the boundary layer, can be calculated in the following way:

$$\theta = \int_0^\delta \frac{\rho(y)}{\rho_0} \cdot \frac{u(y)}{u_0} \cdot \left(1 - \frac{u(y)}{u_0}\right) \cdot dy$$

For incompressible fluids:

$$\theta = \int_0^\delta \frac{u(y)}{u_0} \cdot \left(1 - \frac{u(y)}{u_0}\right) \cdot dy$$

We can use an indicator to know how laminar or turbulent the boundary layer is, using the displacement and impulse thickness.

The shape factor is defined as:

$$H = \frac{\delta^*}{\theta}$$

For a semi-infinite plate, we have these values for each flow:

$$H \sim 2.6 \text{, turbulent flow}$$
$$H \sim 1.3 - 1.4 \text{, laminar flow}$$

The shear force can be expressed as follows:

$$\tau_w = \rho_0 \cdot u_0^2 \cdot \frac{d\theta}{dx}$$

Let's calculate the shear force for a velocity profile:

$$\frac{u}{u_0} = \left(\frac{y}{\delta}\right)^{1/7}$$

Now, we calculate the impulse thickness:

$$\theta = \int_0^\delta \left(\frac{y}{\delta}\right)^{1/7} \cdot \left(1 - \left(\frac{y}{\delta}\right)^{1/7}\right) \cdot dy$$

Solving the integral:

$$\theta = \frac{7}{72} \cdot \delta$$

Therefore:

$$\tau_w = \rho_0 \cdot u_0{}^2 \cdot \frac{7}{72} \cdot \frac{d\delta}{dx}$$

We compare it with the value experimentally obtained:

$$\tau_w = 0.0225 \cdot \rho_0 \cdot u_0{}^2 \cdot \left(\frac{\mu_0}{\rho_0 \cdot u_0 \cdot \delta}\right)^{1/4}$$

If we equal both values, obtain the following differential equation:

$$0.0225 \cdot \left(\frac{\mu_0}{\rho_0 \cdot u_0 \cdot \delta}\right)^{1/4} = \frac{7}{72} \cdot \frac{d\delta}{dx}$$

We solve the equation separating variables:

$$0.231 \cdot \left(\frac{\mu_0}{\rho_0 \cdot u_0}\right)^{1/4} \cdot dx = \delta^{1/4} \cdot d\delta$$

Solving the integral:

$$0.231 \cdot \left(\frac{\mu_0}{\rho_0 \cdot u_0} \right)^{1/4} \cdot x = \frac{4}{5} \cdot \delta^{5/4}$$

We can express the result as follows:

$$\delta = 0.37 \cdot \left(\frac{\mu_0}{\rho_0 \cdot u_0} \right)^{1/5} \cdot x^{4/5}$$

DRAG COEFFICIENT

The resistance or drag is linked to the friction and, therefore, to the viscosity.

The drag coefficient can be defined as:

$$C_D = \frac{Drag}{\frac{1}{2} \cdot \rho_0 \cdot u_0^2 \cdot b \cdot l}$$

It can be represented as a function of the Reynolds number:

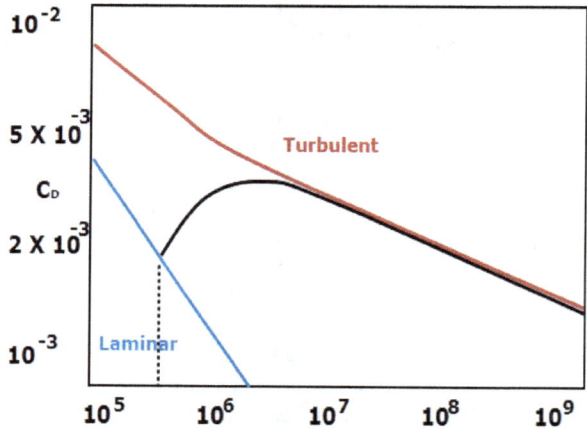

Let's calculate the friction coefficient for the following velocity profile:

$$\frac{u}{u_0} = \left(\frac{y}{\delta}\right)^{1/7}$$

The Drag is determined by:

$$Drag = b \cdot \int_0^1 \tau_w \cdot dx$$

If we place the share force depending on the distance on the other side of the plate, we have the following:

$$\tau_w = 0.0288 \cdot \rho_0 \cdot u_0{}^2 \cdot \left(\frac{\mu_0}{\rho_0 \cdot u_0 \cdot \delta}\right)^{1/5} \cdot x^{-1/5}$$

We have to solve the integral in order to find the *Drag:*

$$Drag = b \cdot \int_0^l 0.0288 \cdot \rho_0 \cdot u_0{}^2 \cdot \left(\frac{\mu_0}{\rho_0 \cdot u_0 \cdot \delta}\right)^{1/5} \cdot x^{-1/5} \cdot dx$$

The obtained result is:

$$Drag = 0.036 \cdot b \cdot \rho_0 \cdot u_0{}^2 \cdot \left(\frac{\mu_0}{\rho_0 \cdot u_0 \cdot \delta}\right)^{1/5} \cdot l^{4/5}$$

If we replace the *Drag* force in the expression (given by the *drag* coefficient), we arrive to the following result:

$$C_D = 0.072 \cdot \left(\frac{\mu_0}{\rho_0 \cdot u_0 \cdot \delta}\right)^{1/5}$$

BLS – BOUNDARY LAYER SUCTION

There are several methods to control attachment of the boundary layer.

Let's suppose that we have a wing with a determined angle of attack. By using wool tufts, we can know the air dynamics on the surface:

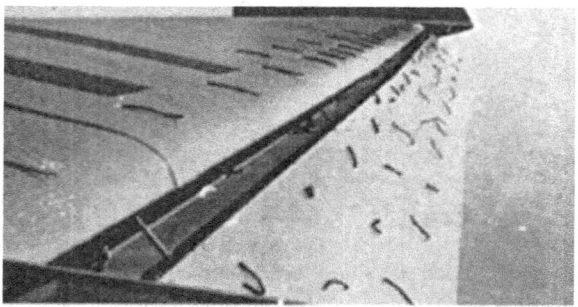

We can see that the air is not attached to the wings. Therefore, the resistance is increased and lift is decreased. To avoid this, some holes are placed at the end of the wing in order to suck out the air flow. Porous materials, in areas where suction is made, can be installed.

We can see that the flow is now attached to the surface. Thanks to this, the air tends to go down and, therefore, the plane tends to go up (the action-reaction law). In the same way and with the same goal, instead of placing the holes at the end of the wing (where the flow is sucked out), we can place the holes (where high velocity air is injected) at the "beginning" of the *flap*-wing. This causes a high velocity movement of molecules on the surface, which drags the main flow on it.

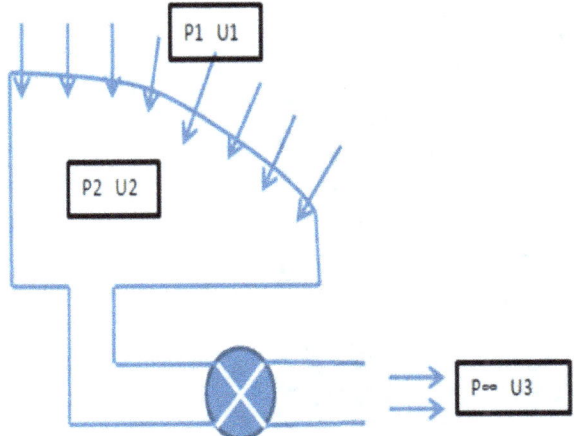

If we increase "excessively" the angle of attack of the front wing (over the limit angle that the car manual indicates), the lift decreases quickly because the boundary layer will detach from the wing.

Let's place the "BLS" (or "Boundary Layer Suction") device on a F3. We will be able to reach higher angles of attack with the same wing, reaching higher lift levels or downforce. Moreover, with lower resistance values and higher top speed.

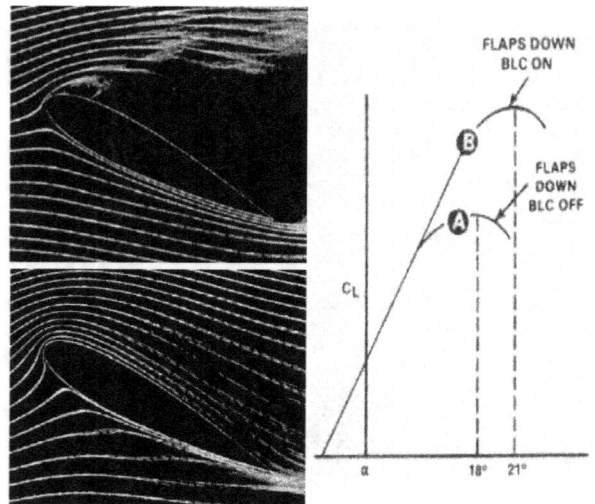

Sample device:

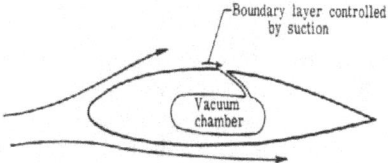

Is possible, in order to avoid the same goal, create a injector fluid:

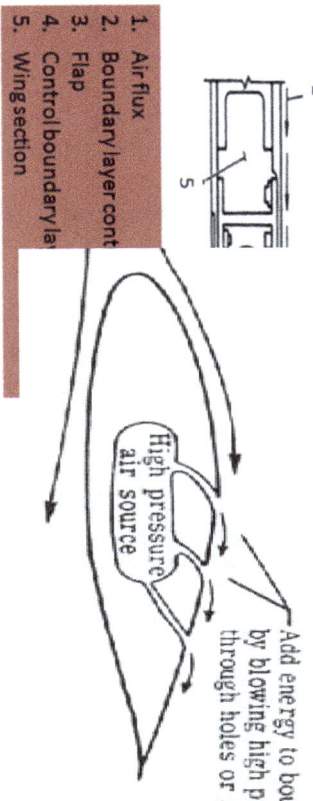

1. Air flux
2. Boundary layer cont[...]
3. Flap
4. Control boundary la[...]
5. Wing section

High pressure air source

Add energy to boundary layer by blowing high pressure air through holes or slots

There are a lot of books that quantify and explain how the evolution of the boundary layer is performed. Prandtl published a lot of information about it, Karman, Blasius and other authors as well. Unfortunately, most of their numeric developments are not very useful in our application field because almost all of them talk about flat "infinite" plates and don't talk about the boundary layer with respect to obstacles and interactions among bodies.

If we manage to avoid the separation of the flow on a body, the boundary layer remains smooth and thin, decreasing the *drag* pressure. It is important to consider the nature of the boundary layer whether it is laminar or turbulent this will make a huge difference on our performance. In this way, a large movement transfer within the turbulent boundary layer requires a larger adverse pressure gradient than in the case of a laminar flow in order to cause the separation. Hence, if the flow is turbulent we can reach higher angles of attack. If the flow is not turbulent, we can make it become turbulent using turbulators.

In other words, there are two options regarding air flow:

- Attaching it to the surface by suction or impulsion.

- Making the flow become turbulent, so that it attaches to the surface on its own.

As it can be imagined, what we want to do is to "sharpen" the rear part of the car so that the flow is always in contact with the object. In this way, the boundary layer will not detach:

For example, we know we have to try and avoid the flux to be detached on a wing; thanks to this we will generate higher downforce; we can generate a depression or suction (Boundary layer Suction — BLS); example; rear part of a car:

As you can see in this picture, the low pressure produced behind the car sucks air from inside the lateral ducts (brakes refrigeration, may be):

If we refine the rear section, we will reduce the drag. In following chapters we will see all the possible options to accomplish this objective, but knowing that one fact is to not separate the boundary layer is enough now:

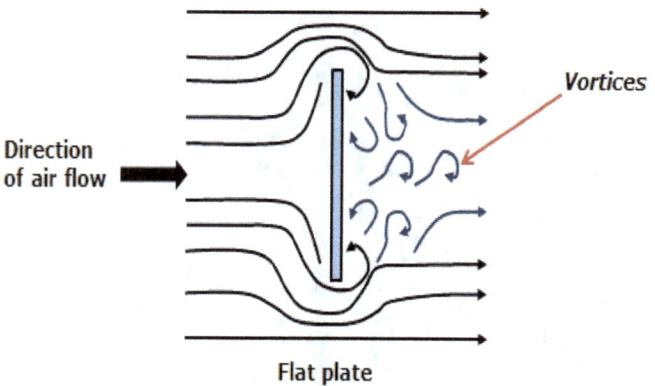

Flat plate

Circular section   Circular / lobe section

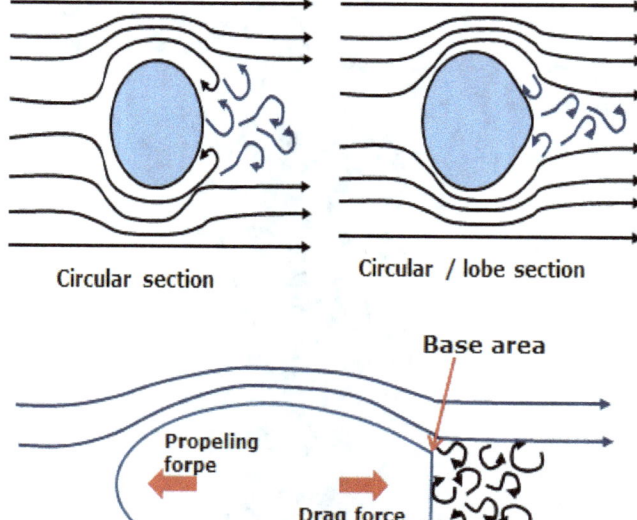

Bottaled rear drop

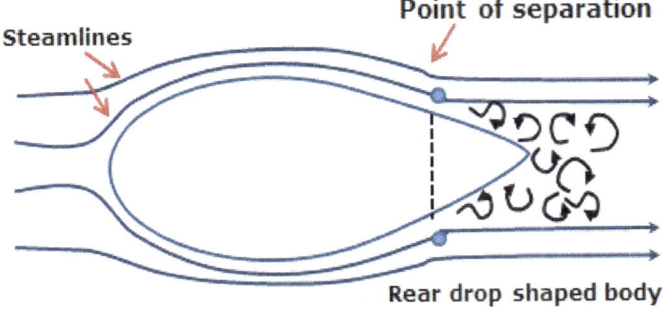

Steamlines

Point of separation

Rear drop shaped body

Red Bull has been using a little crack below the cockpit. This device allows the team to reduce the boundary layer thickness and, therefore, reduce the drag:

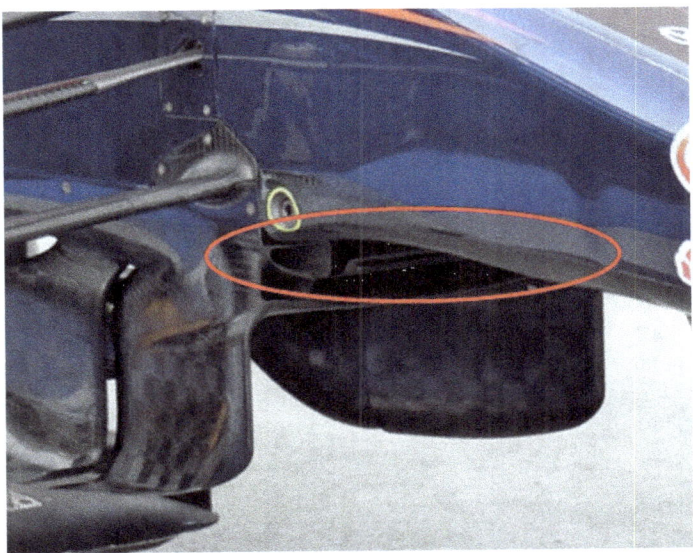

In aircrafts, is possible to see the holes, in order to suck the boundary layer:

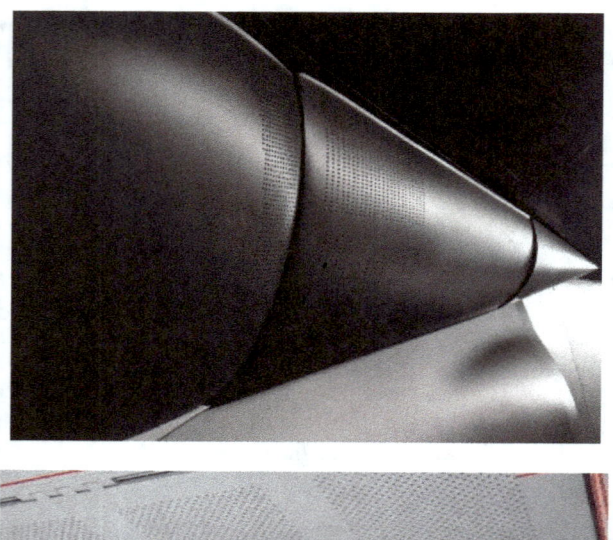

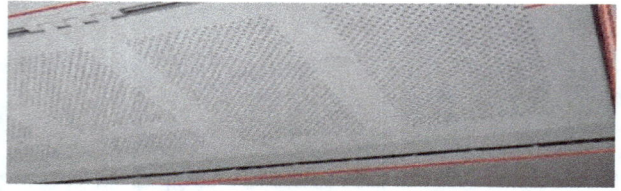

BOUNDARY LAYER EVOLUTION

The mathematical models used to know the thickness of the boundary layer, are based on making different suppositions about the velocity evolution within the boundary layer. We suppose that the velocity evolves as an equation of grade 2, grade 3 or other development evolution. All of them arrive to results like:

$$u(y = 0) = 0$$
$$u(y = \delta) = u_\infty$$
$$\left.\frac{\partial u}{\partial y}\right|_{y=\delta} = 0$$

$$\frac{\partial^2 u}{\partial y^2}\bigg|_{y=0} = 0$$

Laminar:

$$\frac{u}{u_\infty} = A + By + Cy^2 + Dy^3$$

$$\frac{u}{u_\infty} = \frac{3}{2}\left(\frac{y}{\delta}\right) + \frac{1}{2}\left(\frac{y}{\delta}\right)^3$$

Applying the conditions above, we have:
Where: A=C=0

Turbulent:

$$\frac{u}{u_\infty} = \left(\frac{y}{\delta}\right)^{\frac{1}{n}}$$

$$n = 7 \quad Re_x < 10^7$$

$$n = 8 \quad 10^7 < Re_x < 10^8$$

$$n = 9 \quad 10^8 < Re_x < 10^9$$

Obtaining expressions like:

$$\frac{\delta}{x} = \frac{5}{\sqrt{Re_x}}$$

$$\frac{\delta}{x} = \frac{0.385}{Re_x{}^{0.2}}$$

Re_x, denotes the Reynolds number in the point "x".

The first expression belongs to the laminar flow and the second one belongs to the turbulent one.

Within the boundary layer, we can see laminar and turbulent areas, but all equations that describes the thickness, use values that belong to studies of flat plates. The reality is much more complicated. On a racing car, knowing the thickness of the boundary layer is really difficult and complicated. Knowing where or when the flow detaches from the surface is also difficult. In the expressions above, we can see that the turbulent boundary layer is thicker than the laminar one. In any case, we can suppose an "invalid" velocity on a surface, with which we obtain a parabolic velocity profile, applying the Navier Stokes equations:

The Gibbs expression of the N-S equation and the continuity equation are:

Navier Stokes equations in Gibbs format:

$$\rho \cdot \left(\frac{\partial v_x}{\partial t} + v_x \cdot \frac{\partial v_x}{\partial x} + v_y \cdot \frac{\partial v_x}{\partial y} + v_z \cdot \frac{\partial v_x}{\partial z} \right) = -\frac{\partial P}{\partial x} + \mu \cdot \left(\frac{\partial^2 v_x}{\partial x^2} + \frac{\partial^2 v_x}{\partial y^2} + \frac{\partial^2 v_x}{\partial z^2} \right) + \rho \cdot g_x$$

$$\rho \cdot \left(\frac{\partial v_y}{\partial t} + v_x \cdot \frac{\partial v_y}{\partial x} + v_y \cdot \frac{\partial v_y}{\partial y} + v_z \cdot \frac{\partial v_y}{\partial z} \right) = -\frac{\partial P}{\partial y} + \mu \cdot \left(\frac{\partial^2 v_y}{\partial x^2} + \frac{\partial^2 v_y}{\partial y^2} + \frac{\partial^2 v_y}{\partial z^2} \right) + \rho \cdot g_y$$

$$\rho \cdot \left(\frac{\partial v_z}{\partial t} + v_x \cdot \frac{\partial v_z}{\partial x} + v_y \cdot \frac{\partial v_z}{\partial y} + v_z \cdot \frac{\partial v_z}{\partial z} \right) = -\frac{\partial P}{\partial z} + \mu \cdot \left(\frac{\partial^2 v_z}{\partial x^2} + \frac{\partial^2 v_z}{\partial y^2} + \frac{\partial^2 v_z}{\partial z^2} \right) + \rho \cdot g_z$$

Continuity equation:

$$\frac{\partial \rho}{\partial t} + \left(v_x \cdot \frac{\partial \rho}{\partial x} + v_y \cdot \frac{\partial \rho}{\partial y} + v_z \cdot \frac{\partial \rho}{\partial z} \right) + \rho \cdot \left(\frac{\partial v_x}{\partial x} + \frac{\partial v_y}{\partial y} + \frac{\partial v_z}{\partial z} \right) = 0$$

The problem we have is a flat plate with an angle of attack and we have to find the velocity profile on the plate; the height of the boundary layer = H:

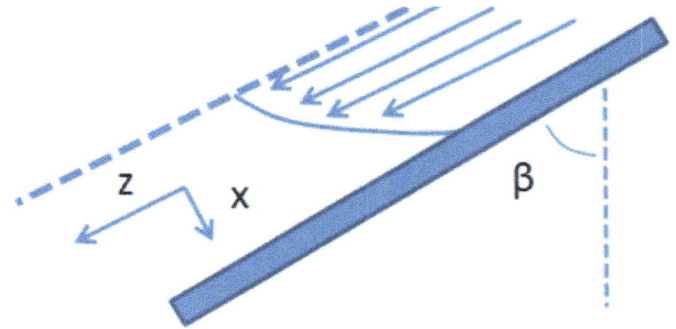

With the problem hypothesis and applying N-S to this particular problem, we obtain the following:

$$\frac{\partial \rho}{\partial t} = \frac{\partial \rho}{\partial x} = \frac{\partial \rho}{\partial y} = \frac{\partial \rho}{\partial z} = 0$$

$$\frac{\partial v_x}{\partial x} = \frac{\partial v_y}{\partial y} = 0$$

$$\rightarrow \frac{\partial v_z}{\partial z} = 0$$

For N-S:

$$v_x = \frac{\partial v_x}{\partial t} = \frac{\partial v_x}{\partial x} = \frac{\partial v_x}{\partial y} = \frac{\partial v_x}{\partial z} = \frac{\partial^2 v_x}{\partial x^2} = \frac{\partial^2 v_x}{\partial y^2} = \frac{\partial^2 v_x}{\partial z^2} = 0$$

$$v_y = \frac{\partial v_y}{\partial t} = \frac{\partial v_y}{\partial y} = \frac{\partial v_y}{\partial z} = \frac{\partial^2 v_y}{\partial x^2} = \frac{\partial^2 v_y}{\partial y^2} = \frac{\partial^2 v_y}{\partial z^2} = 0$$

$$\frac{\partial v_z}{\partial t} = \frac{\partial v_z}{\partial z} = \frac{\partial^2 v_z}{\partial z^2} = 0$$

Replacing these values in N-S:

$$-\frac{\partial P}{\partial x} + \rho \cdot g_x = 0$$

$$-\frac{\partial P}{\partial y} + \rho \cdot g_y = 0$$

$$-\frac{\partial P}{\partial z} + \mu \cdot \left(\frac{\partial^2 v_z}{\partial x^2} + \frac{\partial^2 v_z}{\partial y^2} \right) + \rho \cdot g_z = 0$$

Using geometry, we find the components of the gravitational acceleration:

If we replace:

$$\hat{g} = \begin{pmatrix} g \cdot \sin(\beta) \\ 0 \\ g \cdot \cos(\beta) \end{pmatrix}$$

$$\frac{\partial P}{\partial x} = \rho \cdot g \cdot sin(\beta) \qquad \frac{\partial P}{\partial y} = 0$$

$$\mu \cdot \left(\frac{\partial^2 v_z}{\partial x^2} \right) + \rho \cdot g \cdot cos(\beta) = 0$$

$$\frac{\partial^2 v_z}{\partial x^2} = \frac{-\rho \cdot g \cdot cos(\beta)}{\mu} = a$$

$$v_z = \frac{a \cdot x^2}{2} + c_1 \cdot x + c_2$$

$$\frac{dv}{dx} = a \cdot x + c_1 = 0 \rightarrow c_1 = 0$$

Original condition: vZ=0; x=H. No slip. Original condition in x=0.

$$\frac{a \cdot H^2}{2} + c_1 \cdot H + c_2 = 0$$

Therefore, we find what we were looking for:

$$v_z = \frac{a \cdot x^2}{2} + c_2$$

It is important to know that the laminar boundary layer has less friction than the turbulent one. Although it is not straightly true. Everything depends on the Reynolds number. For a same "Re", where flow is turbulent, friction is higher:

Friction coefficient depends on the Reynolds number:

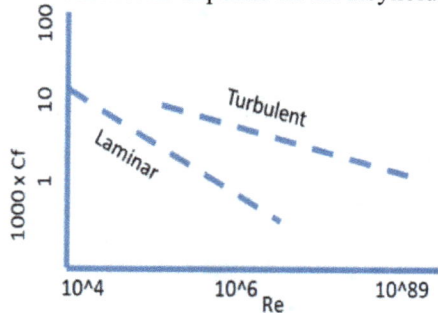

Other examples:

1. Couette Flow:

The Couette Flow is defined as: *"the flow at the space between two parallel plates, one of which is moving relative to the other. The flow is driven by virtue of viscous drag force acting on the fluid and the applied pressure gradient parallel to the plates."*

Set up the equations and boundary conditions to solve for the following problem at steady state and fully developed:

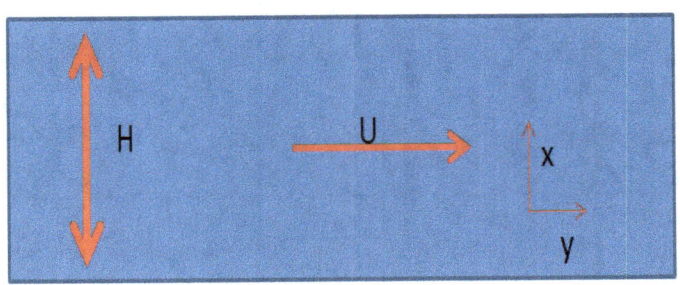

$$\rho g_x - \frac{\partial P}{\partial x} + \mu \left(\frac{\partial^2 v_x}{\partial x^2} + \frac{\partial^2 v_x}{\partial y^2} + \frac{\partial^2 v_x}{\partial z^2} \right) = \rho \left(\frac{\partial v_x}{\partial t} + v_x \frac{\partial v_x}{\partial x} + v_y \frac{\partial v_x}{\partial y} + v_z \frac{\partial v_x}{\partial z} \right)$$

$$\rho g_y - \frac{\partial P}{\partial y} + \mu \left(\frac{\partial^2 v_y}{\partial x^2} + \frac{\partial^2 v_y}{\partial y^2} + \frac{\partial^2 v_y}{\partial z^2} \right) = \rho \left(\frac{\partial v_y}{\partial t} + v_x \frac{\partial v_y}{\partial x} + v_y \frac{\partial v_y}{\partial y} + v_z \frac{\partial v_y}{\partial z} \right)$$

$$\rho g_x - \frac{\partial P}{x} + \mu \left(\frac{\partial^2 v_x}{\partial x^2} + \frac{\partial^2 v_x}{\partial y^2} \right) = \rho \left(v_x \frac{\partial v_x}{\partial x} + v_y \frac{\partial v_x}{\partial y} \right)$$

$$\rho g_y - \frac{\partial P}{\partial y} + \mu \left(\frac{\partial^2 v_y}{\partial x^2} + \frac{\partial^2 v_y}{\partial y^2} \right) = \rho \left(v_x \frac{\partial v_y}{\partial x} + v_y \frac{\partial v_y}{\partial y} \right)$$

H U→ X y

$$-\frac{\vartheta P}{\vartheta x} + \mu \left(\frac{\vartheta^2 v_x}{\vartheta x^2} + \frac{\vartheta^2 v_x}{\vartheta y^2}\right) = \rho \left(v_x \frac{\vartheta v_x}{\vartheta x}\right)$$

$$\rho g_y - \frac{\vartheta P}{\vartheta y}$$

$$\mu \frac{\vartheta^2 v_x}{\vartheta y^2} = \frac{\vartheta P}{\vartheta x} \qquad\qquad \mu \frac{\vartheta^2 v_x}{\vartheta y^2} = 0$$

$$\frac{\vartheta v_x}{\vartheta y} = C_1$$

$$\rho g_y \, \text{y} + c(x) = P(x, y) \qquad v_x = C_1 y + C_2$$

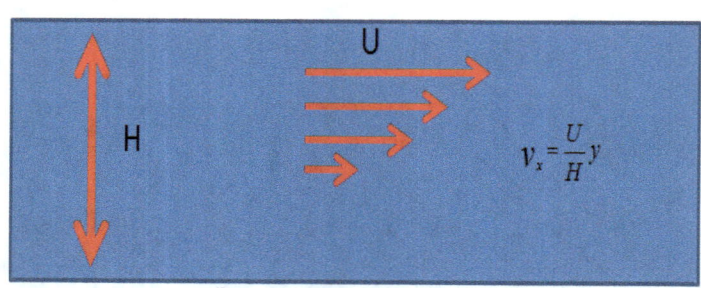

U

H

$$v_x = \frac{U}{H}y$$

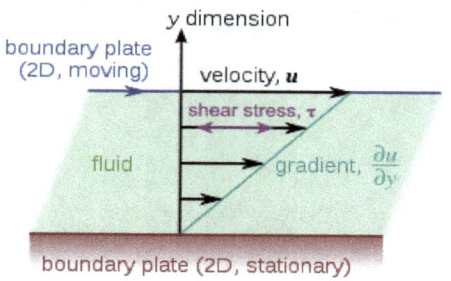

y dimension

boundary plate (2D, moving)

velocity, **u**

shear stress, τ

fluid

gradient, $\frac{\partial u}{\partial y}$

boundary plate (2D, stationary)

2.   Poiseuille Flow:

The Poiseuille flow is defined as: *"Steady viscous fluid flow driven by an effective pressure gradient established between the two ends of a long straight pipe of uniform circular cross-section".*

In order to improve the clarity of the example, we will consider the last problem, but without a moving wall and with a pump providing a pressure gradient dP/dx.

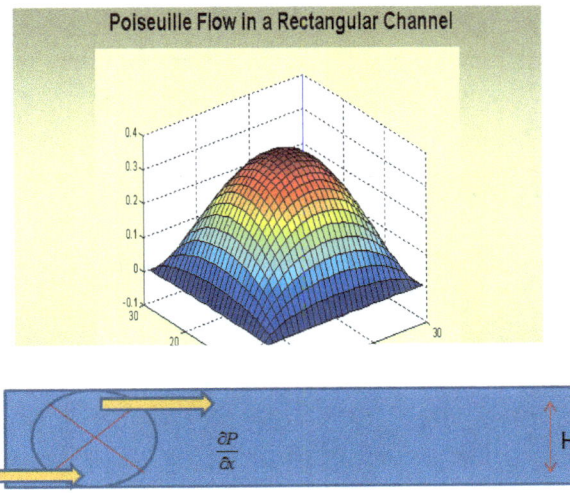

$$\rho g_x - \frac{\partial P}{\partial x} + \mu \left(\frac{\partial^2 v_x}{\partial x^2} + \frac{\partial^2 v_x}{\partial y^2} + \frac{\partial^2 v_x}{\partial z^2} \right) = \rho \left(\frac{\partial v_x}{\partial t} + v_x \frac{\partial v_x}{\partial x} + v_y \frac{\partial v_x}{\partial y} + v_z \frac{\partial v_x}{\partial z} \right)$$

$$\rho g_y - \frac{\partial P}{\partial y} + \mu \left(\frac{\partial^2 v_y}{\partial x^2} + \frac{\partial^2 v_y}{\partial y^2} + \frac{\partial^2 v_y}{\partial z^2} \right) = \rho \left(\frac{\partial v_y}{\partial t} + v_x \frac{\partial v_y}{\partial x} + v_y \frac{\partial v_y}{\partial y} + v_z \frac{\partial v_y}{\partial z} \right)$$

$$\mu \frac{\partial^2 v_x}{\partial y^2} = \frac{\partial P}{\partial x}$$

$$\frac{\partial v_x}{\partial y} = \frac{1}{\mu} \frac{\partial p}{\partial x} y + C_1$$

$$v_x = \frac{1}{2\mu} \frac{\partial p}{\partial x} y^2 + C_1 y + C_2$$

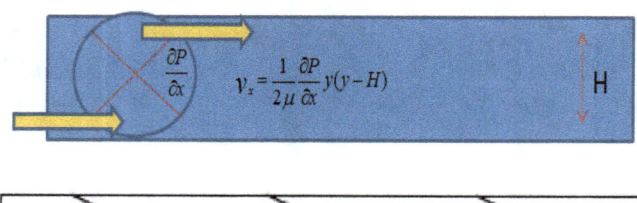

$$v_z = \frac{1}{2\mu} \frac{\partial P}{\partial x} y(y-H)$$

H

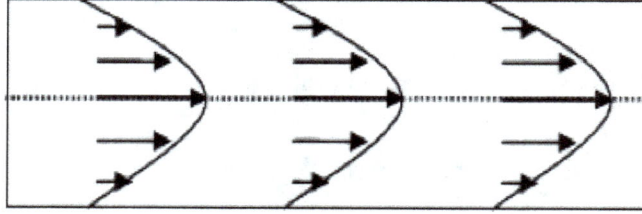

EXAMPLES-1

Let's look at an example for the calculation of a laminar area in a front part of a car: F1 and GT:

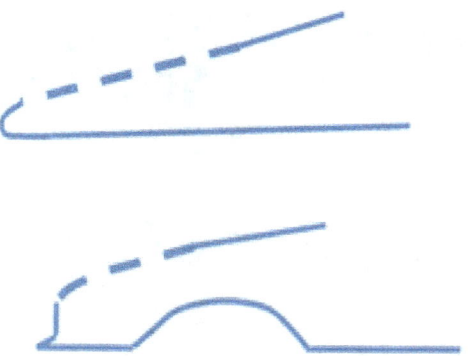

Let's think of a front part of a car. We know that in x = 0.5 m, the thickness of the boundary layer is δ=6 mm. We want to know if the flow is laminar or turbulent and where it will become turbulent:

$$\delta = 5 \cdot \sqrt{\frac{v \cdot x}{V}}$$

V = 0.39 m/s; sup. laminar.

$$Re = \frac{0.39 \cdot 0.5}{1.12 \cdot 10^{-6}} = 1.74 \cdot 10^5 < 5 \cdot 10^5$$

"V" is the velocity and "v" is the kinematic viscosity. As we have previously said, $5 \cdot 10^5$ is the value that points out the change from laminar to turbulent. Where does this take place?:

If Re=$5 \cdot 10^5$:

$$x_{critical} = \frac{5 \cdot 10^5 \cdot 1.12 \cdot 10^{-6}}{0.39} = 1.43 \, m$$

$$\delta = 5 \cdot \sqrt{\frac{v \cdot x}{V}} = 0.01 \, m$$

Now, let's look at another typical example: we have to calculate the height that a pitot tube placed on a F1 should be so that the boundary layer doesn't interfere in the measuring. To do this, we have to calculate the boundary layer height, designing the pitot height with an appropriate safety margin; position "x" = 1 m:

Let's suppose that the air density is 1.225 kg/m^3

Viscosity = $1.8 \cdot 10\text{-}5$ kg/ms

The fastest velocity of the air = 100 m/s; 10 m/s minimum.

Let's suppose that the transition Reynolds number is 500.000

$$\delta_{laminar} = \frac{5 \cdot x}{\sqrt{Re_x}}$$

$$Re_x = \frac{\rho \cdot V \cdot x}{\mu}$$

The Pitot tube has to be placed at 1 m in the front part of the car:

$$\delta_{turbulent} = \frac{0.375 \cdot x}{Re_x^{0.2}}$$

$$x_{transition} = \frac{Re_{transition}}{\frac{\rho \cdot V}{\mu}}$$

$$x_{transition} = \frac{500000}{\frac{1.225 \cdot 100 \cdot 1}{1.8 \cdot 10^{-5}}} = 0.0735\ m$$

At this moment, we already know that the transition point is before the situation of the Pitot tube. Consequently, we will use the equation corresponding to the thickness of the turbulent boundary layer:

$$\delta_{turbulent} = \frac{0.375 \cdot 1}{\sqrt[5]{\frac{1.225 \cdot 100 \cdot 1}{1.8 \cdot 10^{-5}}}} = 0.016\ m$$

Therefore, with an appropriate safety margin, we design the pilot with a height of 5 cm (in contrast with the calculated 1.6 cm).

EXAMPLES-2

Example 1. (Boundary Layer Thickness):

Velocity profiles in laminar boundary layers often are approximated by the equations:

Linear: $\dfrac{u}{U} = \dfrac{y}{\delta}$

Sinusoidal:

$$\frac{u}{U} = \sin\left(\frac{\pi}{2}\frac{y}{\delta}\right)$$

Parabolic:

$$\frac{u}{U} = 2\left(\frac{y}{\delta}\right) - \left(\frac{y}{\delta}\right)^2$$

Calculate δ^* (displacement thickness) and θ (momentum thickness) for these velocity profiles and compare the result for each case.

Statement of the Problem:

Given:

- Three approximated velocity profiles, linear, sinusoidal, and parabolic, in laminar boundary layers.

Find:

- δ^* (displacement thickness) for three approximated velocity profiles.
- θ (momentum thickness) for three approximated velocity profiles.
- Compare the result for each case.

System Diagram:

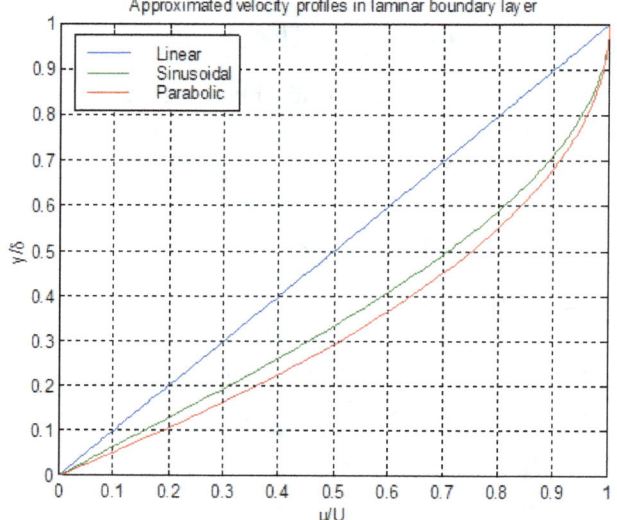

Approximated velocity profiles in laminar boundary layer

Assumptions:

- Steady state condition.
- Laminar boundary layer.

Governing Equations:

- Displacement thickness definition.

$$\delta^* = \int_0^\infty \left(1 - \frac{u}{U}\right) dy \approx \int_0^\delta \left(1 - \frac{u}{U}\right) dy$$

since $u \approx U$ at $y = \delta$, the integrand is essentially zero for $y \geq \delta$.

- Momentum thickness definition:

$$\theta = \int_0^\infty \frac{u}{U} \left(1 - \frac{u}{U}\right) dy \approx \int_0^\delta \frac{u}{U} \left(1 - \frac{u}{U}\right) dy$$

Again, the integrand is essentially zero for $y \geq \delta$.

Detailed Solution: let $\dfrac{y}{\delta} = \eta$, then $dy = \delta \cdot d\eta$ because δ = $\delta(x)$.

δ* (Displacement thickness).

- Linear velocity profile:

$$\delta^*_{linear} = \int_0^\delta \left(1 - \frac{y}{\delta} \right) \cdot dy$$

$$= \int_0^1 (1 - \eta) \cdot \delta \cdot d\eta$$

$$= \delta \cdot \left[\eta - \frac{1}{2} \eta^2 \right]_0^1$$

$$= \frac{1}{2} \delta$$

- Sinusoidal velocity profile:

$$\delta^*_{\sin} = \int_0^\delta \left[1 - \sin\left(\frac{\pi}{2} \frac{y}{\delta} \right) \right] \cdot dy$$

$$= \int_0^1 \left[1 - \sin\left(\frac{\pi}{2} \eta \right) \right] \cdot \delta \cdot d\eta$$

$$= \delta \cdot \left[\eta + \frac{2}{\pi} \cos\left(\frac{\pi}{2} \eta \right) \right]_0^1$$

$$= \left(1 - \frac{2}{\pi} \right) \cdot \delta$$

- Parabolic velocity profile:

$$\delta^*_{parabolic} = \int_0^\delta \left[1 - \left\{ 2\left(\frac{y}{\delta} \right) - \left(\frac{y}{\delta} \right)^2 \right\} \right] \cdot dy$$

$$= \int_0^1 \left[1 - \left(2\eta - \eta^2 \right) \right] \delta \cdot d\eta$$

$$= \delta \cdot \left[\eta - \eta^2 + \frac{1}{3} \eta^3 \right]_0^1$$

$$= \frac{1}{3} \delta$$

θ (Momentum thickness).

- Linear velocity profile:

$$\theta_{linear} = \int_0^\delta \frac{y}{\delta}\left(1 - \frac{y}{\delta}\right) \cdot dy$$

$$= \int_0^1 \eta(1-\eta) \cdot \delta \cdot d\eta$$

$$= \delta \int_0^1 \eta - \eta^2 \cdot d\eta$$

$$= \delta \cdot \left[\frac{1}{2}\eta^2 - \frac{1}{3}\eta^3\right]_0^1$$

$$= \frac{1}{6}\delta$$

- Sinusoidal velocity profile:

$$\theta_{\sin} = \int_0^\delta \sin\left(\frac{\pi}{2}\frac{y}{\delta}\right) \cdot \left[1 - \sin\left(\frac{\pi}{2}\frac{y}{\delta}\right)\right] \cdot dy$$

$$= \int_0^1 \sin\left(\frac{\pi}{2}\eta\right) \cdot \left[1 - \sin\left(\frac{\pi}{2}\eta\right)\right] \cdot \delta \cdot d\eta$$

$$= \delta \cdot \int_0^1 \sin\left(\frac{\pi}{2}\eta\right) - \sin^2\left(\frac{\pi}{2}\eta\right) \cdot d\eta$$

$$= \delta\left[-\frac{2}{\pi}\cos\left(\frac{\pi}{2}\eta\right) - \left\{\frac{\eta}{2} + \frac{\sin\left(2 \cdot \frac{\pi}{2}\eta\right)}{4 \cdot \frac{\pi}{2}}\right\}\right]_0^1$$

$$= \delta \cdot \left(-\frac{1}{2} + \frac{2}{\pi}\right)$$

- Parabolic velocity profile:

$$\theta_{parabolic} = \int_0^\delta \left[2\left(\frac{y}{\delta}\right) - \left(\frac{y}{\delta}\right)^2 \right] \cdot \left[1 - \left\{ 2\left(\frac{y}{\delta}\right) - \left(\frac{y}{\delta}\right)^2 \right\} \right] \cdot dy$$

$$= \int_0^1 \left(2\eta - \eta^2 \right) \left[1 - \left(2\eta - \eta^2 \right) \right] \delta \cdot d\eta$$

$$= \delta \cdot \int_0^1 \left(-\eta^4 + 4\eta^3 - 5\eta^2 + 2\eta \right) d\eta$$

$$= \delta \cdot \left[-\frac{1}{5}\eta^5 + \eta^4 - \frac{5}{3}\eta^3 + \eta^2 \right]_0^1$$

$$= \frac{2}{15}\delta$$

Comparison:

	δ^*/δ	θ/δ
Linear	$1/2 = 50\%$ of B.L.	$1/6 = 16.7\%$ of B.L.
Sinusoidal	$1-2/\pi = 36.3\%$ of B.L.	$-1/2+2/\pi = 13.7\%$ of B.L.
Parabolic	$1/3 = 33.3\%$ of B.L.	$2/15 = 13.3\%$ of B.L.

*** Note: B.L. means "Boundary Layer."

Critical Assessment:

Understand the concepts of displacement thickness (δ^*) and momentum thickness (θ).

This problem illustrates how to calculate them from velocity profiles. The above table shows that $\theta < \delta^* < \delta$ for most types of velocity profiles.

Example 2. (Use of the displacement thickness):

Air flows in the entrance region of a square duct, as shown. The velocity is uniform, $V_1 = 30$ m/s, and the duct is 80 mm square. At a section 0.3 m downstream from the entrance, the displacement thickness, δ^*, on each wall measures 1.0 mm. Determine pressure change between sections ① and ②.

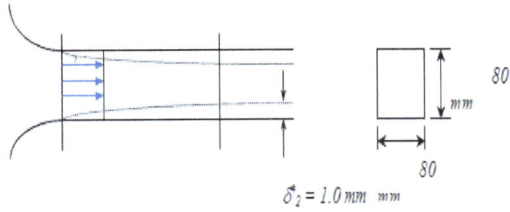

$\delta_2^* = 1.0 \, mm \quad mm$

Statement of the Problem:

Given:

- Working fluid is air which has $\rho_{air} = 1.23 \, kg/m^3$ at $T = 15 \, ℃$.
- Uniform flow at the entrance, $V_1 = 30$ m/s.
- Duct is $H = 80$ mm square.
  - Displacement thickness, $\delta_2^* = 1.0$ mm, on each wall at a section $L = 0.3$ m downstream from the entrance.

Find:

- Pressure change between sections ① and ②.

System Diagram:

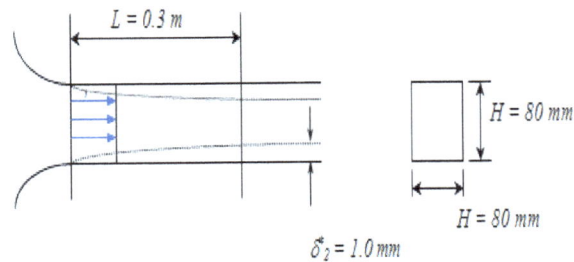

$L = 0.3 \, m$

$H = 80 \, mm$

$H = 80 \, mm$

$\delta_2^* = 1.0 \, mm$

Assumptions:

- Steady state condition.
- Incompressible fluid flow.
- No frictional effects in freestream.
- Flow uniform at each section outside δ^*_2
- Flow along a streamline between sections ① and ②.
- Negligible elevation changes.

Governing Equations:

$$0 = \frac{\partial}{\partial t} \int_{CV} \rho \, dV + \int_{CS} \rho \vec{V} \cdot d\vec{A}$$

... Integral version of mass conservation
Incompressible fluid flow problem, the equation above

$$\Rightarrow 0 = \int_{CS} \vec{V} \cdot d\vec{A}$$

1 inlet (①) and 1 outlet (②) $\Rightarrow$

$$0 = -V_1 A_1 + V_2 A_2$$

- Bernoulli's Equation:

$$\frac{p}{\rho} + \frac{V^2}{2} + gz = const.$$

Restrictions:

(1) Steady flow
(2) Incompressible flow
(3) Frictionless flow
(4) Flow along a streamline

Detailed Solution:

Use the displacement-thickness concept to find the effective flow area for the freestream flow outside the thin wall boundary layers. Replace the actual boundary-layer velocity profiles with uniform velocity profiles as sketched in the following figures.

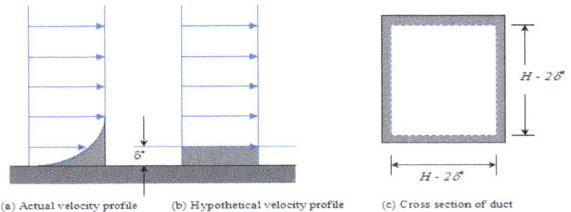

(a) Actual velocity profile   (b) Hypothetical velocity profile   (c) Cross section of duct

Apply the continuity and Bernoulli equations to freestream flow outside the boundary-layer displacement thickness, where viscous effects are negligible.
From Bernoulli's equation, we obtain:

$$\frac{p_1}{\rho} + \frac{V_1}{2} + \cancel{gz_1} = \frac{p_2}{\rho} + \frac{V_2}{2} + \cancel{gz_2}$$

$$p_1 - p_2 = \frac{1}{2}\rho\left(V_2^2 - V_1^2\right)$$

From the continuity equation, we have

$$0 = -V_1A_1 + V_2A_2 \Rightarrow V_2 = \frac{A_1}{A_2}V_1$$

Substituting this expression into Bernoulli's equation,

$$p_1 - p_2 = \frac{1}{2}\rho\left[\left(\frac{A_1}{A_2}V_1\right)^2 - V_1^2\right] = \frac{1}{2}\rho V_1^2\left[\left(\frac{A_1}{A_2}\right)^2 - 1\right]$$

Areas, A_1 and A_2, are:

$$A_1 = H^2$$

$$A_2 = \left(H - 2\delta^*\right)^2 \quad \text{... (only effective}$$

flow area)

Thus,

$$p_1 - p_2 = \frac{1}{2}\rho V_1^2 \left[\left\{ \frac{\left(H^2\right)}{\left(H - 2\delta^*\right)^2} \right\}^2 - 1 \right]$$

After plugging in values,

$$p_1 - p_2 = 58.99(Pa)$$

Critical Assessment:

To solve this problem, it is critical to understand the meaning and physical interpretation of displacement thickness concept.

Example 3. (Use of the Momentum Integral Method):

The velocity profile in a laminar boundary-layer flow at zero pressure gradient is approximated by the linear expression:

$$\frac{u}{U} = \frac{y}{\delta}$$. Use the momentum integral equation with

this profile to obtain expressions for δ/x and C_f.

Statement of the Problem:

Given:

- Laminar boundary-layer flow.
- Zero pressure gradient.
- Velocity profile is approximated by the linear expression, $\dfrac{u}{U} = \dfrac{y}{\delta} = \eta$

Find:

Using the momentum integral equation, obtain expression for

- δ / x
- C_f

System Diagram:

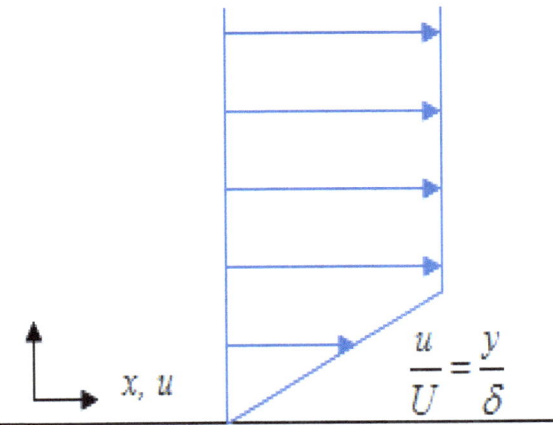

$$\frac{u}{U} = \frac{y}{\delta}$$

x, u

Assumptions:

- Steady state condition.
- Incompressible fluid flow.

Governing Equations:

- Momentum integral equation:

$$\frac{\tau_w}{\rho} = \frac{d}{dx}\left(U^2\theta\right) + \delta^* U \frac{dU}{dx}$$

Where:

$$\delta^* = \int_0^\delta \rho \left(1 - \frac{u}{U}\right) \cdot dy \quad \dots$$

displacement thickness

$$\theta = \int_0^\delta \rho \frac{u}{U} \left(1 - \frac{u}{U}\right) \cdot dy$$

... momentum thickness

Detailed Solution:

For the special case of flow over a flat plate, $U = constant$. From Bernoulli's equation, we see that for this case, $p = constant$, and thus $dp/dx = 0$.

The momentum integral equation then reduces to:

$$\tau_w = \rho U^2 \frac{d\theta}{dx} = \rho U^2 \frac{d}{dx} \int_0^\delta \frac{u}{U}\left(1 - \frac{u}{U}\right) \cdot dy$$

Define $\dfrac{u}{U} = \dfrac{y}{\delta} = \eta \Rightarrow y = \delta\eta \Rightarrow dy = \delta \cdot d\eta$

$(\because \delta = \delta(x))$

$$\therefore \tau_w = \rho U^2 \frac{d}{dx} \int_0^\delta \frac{u}{U}\left(1 - \frac{u}{U}\right) \cdot dy$$

$$= \rho U^2 \frac{d}{dx} \int_0^1 \eta(1-\eta) \cdot \delta \cdot d\eta$$

$$= \rho U^2 \frac{d\delta}{dx} \int_0^1 \left(\eta - \eta^2\right) d\eta$$

$$= \rho U^2 \frac{d\delta}{dx} \left[\frac{1}{2}\eta^2 - \frac{1}{3}\eta^3\right]_0^1$$

$$= \rho U^2 \frac{d\delta}{dx} \left[\left(\frac{1}{2} - \frac{1}{3}\right) - (0 - 0)\right]$$

$$\Rightarrow \quad \tau_w = \frac{1}{6}\rho U^2 \frac{d\delta}{dx}$$

On the other hand, the shear stress can be calculated by:

$$\tau_w = \mu \frac{\partial u}{\partial y}\bigg|_{y=0}$$

And $\dfrac{u}{U} = \dfrac{y}{\delta} \Rightarrow u = \dfrac{U}{\delta} y$, thus:

$$\tau_w = \mu \dfrac{\partial}{\partial y}\left(\dfrac{U}{\delta} y\right)\Bigg|_{y=0}$$

$$= \mu \dfrac{U}{\delta}\dfrac{\partial}{\partial y}(y)\Bigg|_{y=0}$$

$$= \mu \dfrac{U}{\delta}(1)\Bigg|_{y=0}$$

$$\therefore \tau_w = \mu \dfrac{U}{\delta}$$

Comparing (equating) this shear stress equation with the previous shear stress expression:

$$\tau_W = \mu \dfrac{U}{\delta} = \dfrac{1}{6}\rho U^2 \dfrac{d\delta}{dx}$$

$$\frac{d\delta}{dx} = \frac{6\mu}{\rho U} \cdot \frac{1}{\delta}$$

$$\delta \cdot d\delta = \frac{6\mu}{\rho U} \cdot dx$$

$$\int \delta \cdot d\delta = \frac{6\mu}{\rho U} \int dx$$

$$\frac{1}{2}\delta^2 = \frac{6\mu}{\rho U} \cdot x + const$$

When $x = 0$, $\delta = 0 \Rightarrow const = 0$

$$\delta^2 = \frac{12\mu}{\rho U} \cdot x = \frac{12\mu}{\rho U} \cdot \frac{x^2}{x}$$

$$\delta = \sqrt{\frac{12\mu}{\rho Ux}} \cdot x \quad (\because \delta \geq 0)$$

$$\therefore \quad \frac{\delta}{x} = \sqrt{\frac{12}{\frac{\rho U x}{\mu}}} = \frac{\sqrt{12}}{\sqrt{\text{Re}_x}} = \frac{3.46}{\sqrt{\text{Re}_x}}$$

Skin friction coefficient is defined as:

$$C_f = \frac{\tau_w}{\frac{1}{2}\rho U^2}$$

Using the result obtained above:

$$C_f = \frac{\frac{1}{6}\rho U^2 \frac{d\delta}{dx}}{\frac{1}{2}\rho U^2}$$

$$= \frac{1}{3}\frac{d\delta}{dx}$$

$$= \frac{1}{3}\frac{d}{dx}\left[\sqrt{\frac{12\mu}{\rho U}} \cdot \sqrt{x}\right]$$

$$= \frac{1}{3}\sqrt{\frac{12\mu}{\rho U}} \cdot \frac{1}{2} \cdot \frac{1}{\sqrt{x}}$$

$$= \frac{1}{6}\sqrt{\frac{12\mu}{\rho Ux}}$$

$$= \frac{1}{6}\sqrt{\frac{12}{\dfrac{\rho Ux}{\mu}}}$$

$$\therefore C_f = \frac{\sqrt{12}}{6} \cdot \frac{1}{\sqrt{Re_x}} = \frac{0.577}{\sqrt{Re_x}}$$

Critical Assessment:

This problem dealt with linear velocity profile as an approximate solution. The results obtained are rough. However the exercise illustrates the use of the momentum integral method. Practice this method with other types of approximated velocity profile, such as parabolic, sinusoidal, ... etc.

Example 4. (Friction Drag Calculation):

Water at *15* °C flows over a flat plate at a speed of *1 m/s*. The plate is *0.4 m* long and *1 m* wide. The boundary layer on each surface of the plate is laminar. Assume that the velocity profile may be approximated as linear. Determine the drag force on the plate.

Statement of the Problem:

Given:

- Working fluid is water at $T = 15$ °C $\Rightarrow \rho = 999$ *kg/m³* & $\mu = 1.14 \times 10^{-3}$ *N·s/m²*
- $U = 1$ *m/s*

- $L = 0.4\ m$
- $W = 1\ m$
- The boundary layer on each surface of the plate is laminar
- Velocity profile is linear (assuming approximately).

Find:

- Drag force on the plate.

System Diagram:

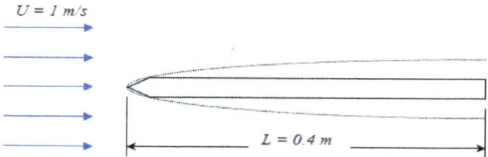

Assumptions:

- Steady state condition.
- Incompressible fluid flow.
- Laminar boundary layer.

Governing Equations:
- Skin friction coefficient definition:

$$C_f = \frac{\tau_w}{\dfrac{1}{2}\rho U^2}$$

- Reynolds number definition for a flat plate:

$$\mathrm{Re}_x = \frac{\rho U x}{\mu}$$

Detailed Solution:

We know that for a linear velocity profile:

$$\frac{u}{U} = \frac{y}{\delta} = \eta$$

$$C_f = \frac{0.577}{\sqrt{\text{Re}_x}}$$

Equating this result and the definition of skin friction coefficient,

$$C_f = \frac{0.577}{\sqrt{\text{Re}_x}} = \frac{\tau_w}{\frac{1}{2}\rho U^2} \Rightarrow$$

$$\tau_w = \frac{1}{2}\rho U^2 \cdot \frac{0.577}{\sqrt{\text{Re}_x}}$$

$$\therefore \tau_w = \frac{1}{2}\rho U^2 \cdot \frac{0.577}{\sqrt{\frac{\rho U x}{\mu}}}$$

Drag force on one side of the plate is given by

$$F_D = \int_{A_p} \tau_w \cdot dA$$

Since $dA = w \cdot dx$ and $0 \leq x \leq L$,

$$F_D = \int_{A_p} \tau_w \cdot dA = \int_0^L \tau_w \cdot w \cdot dx$$

$$F_D = \int_0^L \frac{1}{2} \rho U^2 \cdot \frac{0.577}{\sqrt{\dfrac{\rho U x}{\mu}}} \cdot w \cdot dx$$

$$= \frac{0.577}{2} \rho U^2 w \frac{1}{\sqrt{\dfrac{\rho x}{\mu}}} \int_0^L \frac{1}{\sqrt{x}} \cdot dx$$

$$= \frac{0.577}{2} \rho U^2 w \frac{1}{\sqrt{\dfrac{\rho x}{\mu}}} \cdot \left[2x^{\frac{1}{2}} \right]_0^L$$

$$\therefore F_D = \frac{0.577}{2} \rho U^2 w \frac{1}{\sqrt{\dfrac{\rho x}{\mu}}} \cdot 2\sqrt{L}$$

Plug in values into this expression obtained above for F_D, $\Rightarrow F_D = 0.3894$ N.

For both sides of the plate $\Rightarrow$

$$F_{D,Total} = 2F_D = 0.779$$

N.

Critical Assessment:

Problem says, "the boundary layer on each surface of the plate is laminar." Let us double check that this is true.

$$Re_L = \sqrt{\frac{\rho UL}{\mu}} = 592.05 \quad <<$$

$500,000 = Re_{critical} \Rightarrow$ Obviously it is a laminar flow.

(Note: This problem could be solved by first obtaining the Overall Skin Friction Coefficient, $\overline{C}_f$. In that case, the calculation will proceed by obtaining $\overline{C}_f = \frac{1}{L}\int C_f(x)dx$, where the integration limits will be set at x = 0 and x = L.

$$\text{Then } \overline{C}_f \Rightarrow \overline{\tau}_w = \frac{1}{2}\rho U^2.\overline{C}_f$$

$$\Rightarrow F_D = 2\overline{\tau}_w.A_w \text{ where, } A_w (=W.L)$$

indicates the wet area on each face of the plate).

Example 5. (Power Calculation using Friction Drag):

A flat-bottomed barge, 25 m long and 10 m wide, submerged to a depth of 1.5 m, is to be pushed up a river at the rate of 8 km/hr. Estimate the power required to overcome skin friction if the water temperature is 15 °C.

Statement of the Problem:

Given:

- $L = 25\ m$
- $W = 10\ m$
- $D = 1.5\ m$
- $V = 8\ km/h = 2.222\ m/s$
- Working fluid is water at $T = 15\ °C \Rightarrow \rho = 999$ kg/m^3 & $\mu = 1.14 \times 10^{-3}\ N{\cdot}s/m^2$

Find:

- Power required to overcome skin friction.

System Diagram:

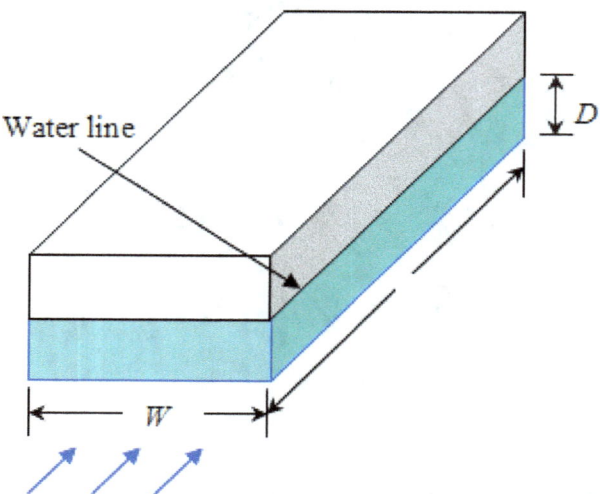

Assumptions:

- Model a flat-bottomed barge as a flat plate.
- Steady state condition.
- Incompressible fluid flow.
- Neglect separation.

Governing Equations:

- Drag Coefficient Definition:

$$C_D = \frac{F_D}{\frac{1}{2}\rho V^2 A}$$

Detailed Solution:

First of all, calculate Reynolds number:

$$\mathrm{Re}_L = \frac{\rho VL}{\mu} = \frac{\left(999 kg/m^3\right)\left(2.222 m/s\right)\left(25m\right)}{\left(1.14\times10^{-3} N\cdot s/m^2\right)} = 4.87\times10^7$$

Transition of laminar to turbulent flow occurs at:

$$\mathrm{Re}_{cr} = 5\times10^5 = \frac{\rho V x_{cr}}{\mu}$$

$$\Rightarrow x_{cr} = \frac{\left(5\times10^5\right)\mu}{\rho V} = \frac{\left(5\times10^5\right)\left(1.14\times10^{-3} N\cdot s/m^2\right)}{\left(999 kg/m^3\right)\left(2.222 m/s\right)} = 0.25678m$$

$$\ll 25\ m$$

This x_{cr} shows that the effect of laminar flow is negligible. It can be said that the flow is turbulent from the leading edge.

For $\mathrm{Re}_L < 10^9$, the empirical equation given by Schlichting:

$$C_D = \frac{0.455}{\left(\log \mathrm{Re}_L\right)^{2.58}}$$

fits experimental data very well.

Friction force is (from the definition of drag coefficient):

$$F_D = \frac{1}{2}\rho V^2 A \cdot C_D = \frac{1}{2}\rho V^2 A \cdot \frac{0.455}{\left(\log \mathrm{Re}_L\right)^{2.58}}$$

$$Power = F_D \cdot V = \frac{1}{2}\rho V^2 A \cdot \frac{0.455}{\left(\log \mathrm{Re}_L\right)^{2.58}} \cdot V \text{ , where A}$$

is the wetted area:

$$A = L \cdot W + 2(L \cdot D)$$

Finally:

$$Power = \frac{1}{2}\rho V^2 \left[L \cdot W + 2(L \cdot D)\right] \cdot \frac{0.455}{\left(\log \mathrm{Re}_L\right)^{2.58}} \cdot V$$

Plug in values into this expression $\Rightarrow \wp = 4200.8 \ W = 4.20 \ Kw$

Critical Assessment:

Drag coefficient must be chosen depending upon the value of Reynolds number for a particular flow condition. Some of C_D expressions are derived by analytical calculation, and others are empirical formulas.

Example 6. (Flow Separation Characteristics):

Two hypothetical boundary-layer velocity profiles are shown. Obtain an expression for the momentum flux of each profile. If the two profiles were subjected to the same pressure gradient conditions, which would be more likely to separate first? Why?

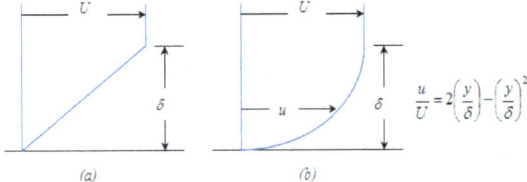

Statement of the Problem:

Given:

- Two hypothetical boundary-layer velocity profiles.

Find:

- Expression for the momentum flux of each profile.
- Which would be more likely to separate first if the two profiles were subjected to the same pressure gradient conditions? And why?.

System Diagram:

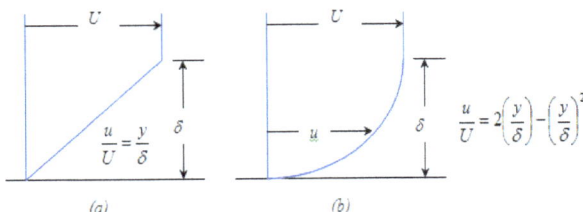

Assumptions:

- Steady state condition.
- Incompressible fluid flow.

Governing Equations:

- Definition of Momentum Flux (mf).
-

$$d\left(mf\right)= \vec{V} \cdot \rho\vec{V} \cdot d\vec{A}$$

Detailed Solution:

Since the flow is 1 - D (positive x direction) and $dA = w \cdot dy$, the momentum equation can be written as:

$$d(mf) = u \cdot \rho \cdot u \cdot dA = u \cdot \rho \cdot u \cdot w \cdot dy$$

$$\therefore \; mf = \int_0^\infty u \cdot \rho \cdot u \cdot w \cdot dy = \int_0^\delta u \cdot \rho \cdot u \cdot w \cdot dy$$

The integrand is essentially zero for $y \geq \delta$.

Linear Velocity Profile:

$$\frac{u}{U} = \frac{y}{\delta} \Rightarrow u = \frac{U}{\delta} y$$

$$mf_{linear} = \int_0^\delta \left(\frac{U}{\delta} y \right) \cdot \rho \cdot \left(\frac{U}{\delta} y \right) \cdot w \cdot dy$$

$$= \rho \frac{U^2}{\delta^2} w \int_0^\delta y^2 \cdot dy$$

$$= \rho \frac{U^2}{\delta^2} w \left[\frac{1}{3} y^3 \right]_0^\delta$$

Finally, $$mf_{linear} = \frac{\rho U^2 w \delta}{3}$$

Parabolic Velocity Profile:

$$\frac{u}{U} = 2\left(\frac{y}{\delta}\right) - \left(\frac{y}{\delta}\right)^2$$

Let $\eta = \dfrac{y}{\delta}$ Then $\dfrac{u}{U} = 2\eta - \eta^2$ and

$$d\eta = \frac{1}{\delta}dy \quad \delta = \delta(x)$$

Now,

$$mf_{parabolic} = \int_0^\delta u \cdot \rho \cdot u \cdot w \cdot dy$$

$$= \int_0^1 u \cdot \rho \cdot u \cdot w \cdot \delta \cdot d\eta$$

$$= \rho \cdot w \cdot \delta \cdot \int_0^1 u^2 \cdot d\eta$$

$$= \rho \cdot w \cdot \delta \cdot \int_0^1 \left[U\left(2\eta - \eta^2\right)\right] d\eta$$

$$= \rho \cdot w \cdot \delta \cdot U^2 \cdot \int_0^1 4\eta^2 - 4\eta^3 + \eta^4 d\eta$$

$$= \rho \cdot w \cdot \delta \cdot U^2 \cdot \left[\frac{4}{3}\eta^3 - \eta^4 + \frac{1}{5}\eta^5\right]_0^1$$

Finally:

$$mf_{parabolic} = \frac{8\rho U^2 w\delta}{15}$$

Which separates first?

Separation occurs when the momentum of fluid layers near the surface is reduced to zero by the combined action of pressure and viscous forces.

As shown in this figure below, the momentum of the fluid near the surface is greater for the parabolic velocity profile.

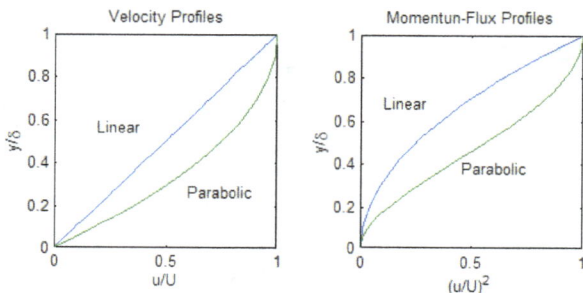

Our previous calculation also shows

$$mf_{linear} = \frac{\rho U^2 w \delta}{3} \quad <$$

$$mf_{parabolic} = \frac{8 \rho U^2 w \delta}{15}$$

Consequently, the parabolic velocity profile is better able to resist separation in the same pressure gradient condition.

$\Rightarrow$ Linear velocity profile would separate first.

Critical Assessment:

Review and understand how the flow separation occurs. Flow separation occurs only when there exists an adverse pressure gradient.

Example 7. (Terminal Velocity Calculation):

A small sphere ($D = 6$ mm) is observed to fall through caster oil at a terminal speed of *60 mm/s*. The temperature is *20 °C*. Compute the drag coefficient for the sphere. Determine the density of the sphere. If dropped in water, would the sphere fall slower or faster? Why?

Statement of the Problem:

Given:

- $D = 6$ mm $= 0.006$ m.
- Working fluid is caster oil at $T = 20$ °C $\Rightarrow S.G._{oil} = 0.969$ & $\mu_{oil} = 0.9$ N·s/m²
- $V_t = 60$ mm/s $= 0.06$ m/s.

Find:

- Drag coefficient for the sphere.
- Density of the sphere.
- If dropped in water, would the sphere fall slower or faster? Why?.

System Diagram:

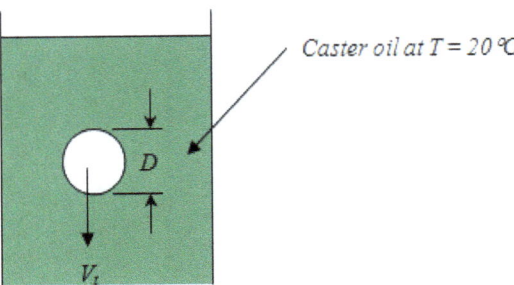

Caster oil at $T = 20$ °C

Assumptions:

- Steady state condition.

- Incompressible fluid flow.

Governing Equations:

- Drag Coefficient Definition:
$$C_D = \frac{F_D}{\frac{1}{2}\rho V^2 A}$$

- Newton's Second Law:
$$\frac{d\vec{P}}{dt} = \sum \vec{F} \,,$$

where $\vec{P}$ is momentum.

When the mass is constant, $m\vec{a} = \sum \vec{F}$

$\Rightarrow$ 1 - D in y direction $\Rightarrow$ $ma_y = \sum F_y$

- Reynolds Number for Sphere: $\mathrm{Re}_D = \frac{\rho V D}{\mu}$

Detailed Solution:

Drag Coefficient:

First of all, calculate Reynolds number:

$$\text{Re}_D = \frac{\rho_{oil} V_t D}{\mu} = \frac{(S.G._{oil} \cdot \rho_{water}) \cdot V_t D}{\mu} =$$

$$\frac{(0.969)(1000 kg / m^3)(0.06 m / s)(0.006 m)}{(0.9 N \cdot s / m^2)}$$

$Re_D = 0.3876 < 1 \Rightarrow$ There is no flow separation from a sphere. The wake is laminar and the drag is predominantly friction drag.

Stokes has shown analytically, for very low Reynolds number flows where inertia forces may be neglected, that drag force on a sphere of diameter D, moving speed V, through a fluid of viscosity μ, is given by:

$$F_D = 3\pi\mu VD$$

The drag coefficient, C_D, is then

$$C_D = \frac{F_D}{\frac{1}{2}\rho V^2 A} = \frac{3\pi\mu VD}{\frac{1}{2}\rho V^2 \left[\frac{\pi}{4}D^2\right]} = \frac{24}{\frac{\rho VD}{\mu}} = \frac{24}{\text{Re}_D}$$

(Note: For sphere, the area, A, is just a cross-sectional area,

which is $\frac{\pi}{4}D^2$.)

Thus,

$$C_D = \frac{24}{\text{Re}_D} = \frac{24}{0.3876} = 61.92$$

Density of the Sphere:

Free Body Diagram:

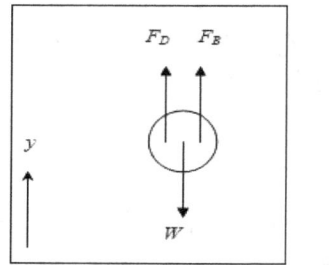

$$W = mg = \rho_s V_s g$$

$$F_D = 3\pi \mu_{oil} V_t D$$

$$F_B = \underset{displaced}{W_{oil}} = \rho_{oil} V_s g$$

The sphere reached the terminal speed $\Rightarrow a_y = 0$

$$ma_y = \sum F_y \quad \text{so,}$$

$$0 = F_D + F_B - W$$

$$0 = 3\pi \mu_{oil} V_t D + \rho_{oil} V_s g - \rho_s V_s g$$

$$\rho_s = \frac{3\pi \mu_{oil} V_t D + \rho_{oil} V_s g}{V_s g}$$

$$\rho_s = \frac{3\pi \mu_{oil} V_t D + (S.G._{oil} \cdot \rho_{water}) \cdot \left[\frac{4}{3}\pi \left(\frac{D}{2}\right)^3 \right] \cdot g}{\left[\frac{4}{3}\pi \left(\frac{D}{2}\right)^3 \right] \cdot g}$$

After plug in values into this expression, $\rho_s = 3721$ kg/m^3.

If dropped in water ...

If the working fluid is water at $T = 20$ °C $\Rightarrow \rho_w = 998$ kg/m³ & $\mu_w = 1 \times 10^{-3}$ N·s/m²

Because $\mu_w = 1 \times 10^{-3}$ N·s/m² $\ll \mu_{oil} = 0.9$ N·s/m², the author guesses the sphere drops faster in water than in caster oil.

If it's faster and $\mu_w \ll \mu_{oil}$,

$$\mathrm{Re}_{D \atop water} \gg \mathrm{Re}_{D \atop oil} = 0.3876$$

$\Rightarrow F_D = 3\pi\mu VD$ cannot be used because the equation works only for very low Reynolds number which we don't know whether this is appropriate or not any more for this case.

Now, guess a value of C_D (Drag coefficient of a smooth sphere as a function of Reynolds number) and calculate V_{tW}. Then calculate Re_D and verify the chosen C_D was appropriate or not.

Guess $C_D = 0.4 \Rightarrow$

$$F_D = \frac{1}{2}\rho_w V_{tW}^2 A \cdot C_D$$

Free Body Diagram (again):

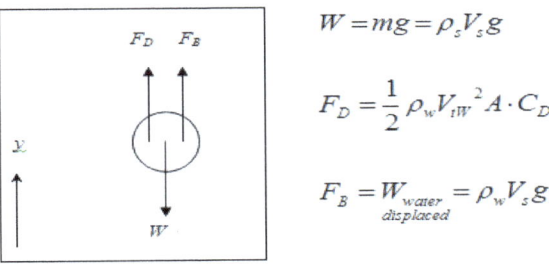

$$W = mg = \rho_s V_s g$$

$$F_D = \frac{1}{2}\rho_w V_{tW}^2 A \cdot C_D$$

$$F_B = W_{water \atop displaced} = \rho_w V_s g$$

The sphere reaches a new terminal speed, $V_{tW} \Rightarrow a_y = 0$

$$ma_y = \sum F_y$$

$$\Downarrow$$

$$0 = F_D + F_B - W$$

$$0 = \frac{1}{2}\rho_w V_{tW}^2 A \cdot C_D + \rho_w V_s g - \rho_s V_s g$$

$$V_{tW} = \sqrt{\frac{(\rho_s - \rho_w)V_s g}{\frac{1}{2}\rho_w A C_D}} = \sqrt{\frac{(\rho_s - \rho_w)\cdot\left[\frac{4}{3}\pi\left(\frac{D}{2}\right)^3\right]\cdot g}{\frac{1}{2}\rho_w\left[\frac{\pi}{4}D^2\right]\cdot C_D}}$$

After plugging values into this expression, $V_{tW} = 0.732$ m/s.

With this new terminal speed, Reynolds number is:

$$\text{Re}_D = \frac{\rho_w V_{tW} D}{\mu_w} = 4383$$

When $C_D = 0.4$, $Re_D \approx 4 \times 10^3$, which is about right for this case. This shows the new terminal speed is a valid number.

$V_{tW} = 0.732 \ m/s \ > \ V_t = 0.06 \ m/s \Rightarrow$ The sphere drops faster in water than in caster oil.

Critical Assessment:

Drag coefficient depends upon the value of Reynolds number. Be careful with choosing a right C_D depending on a particular flow condition.

More problems about:

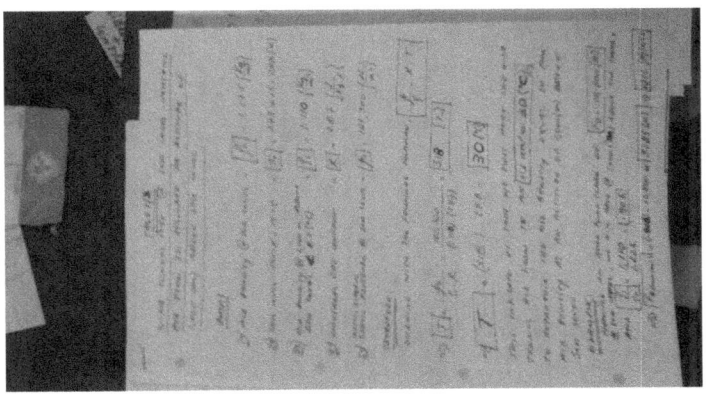

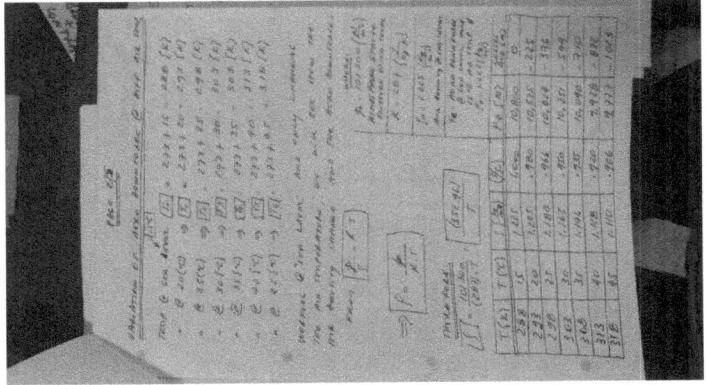

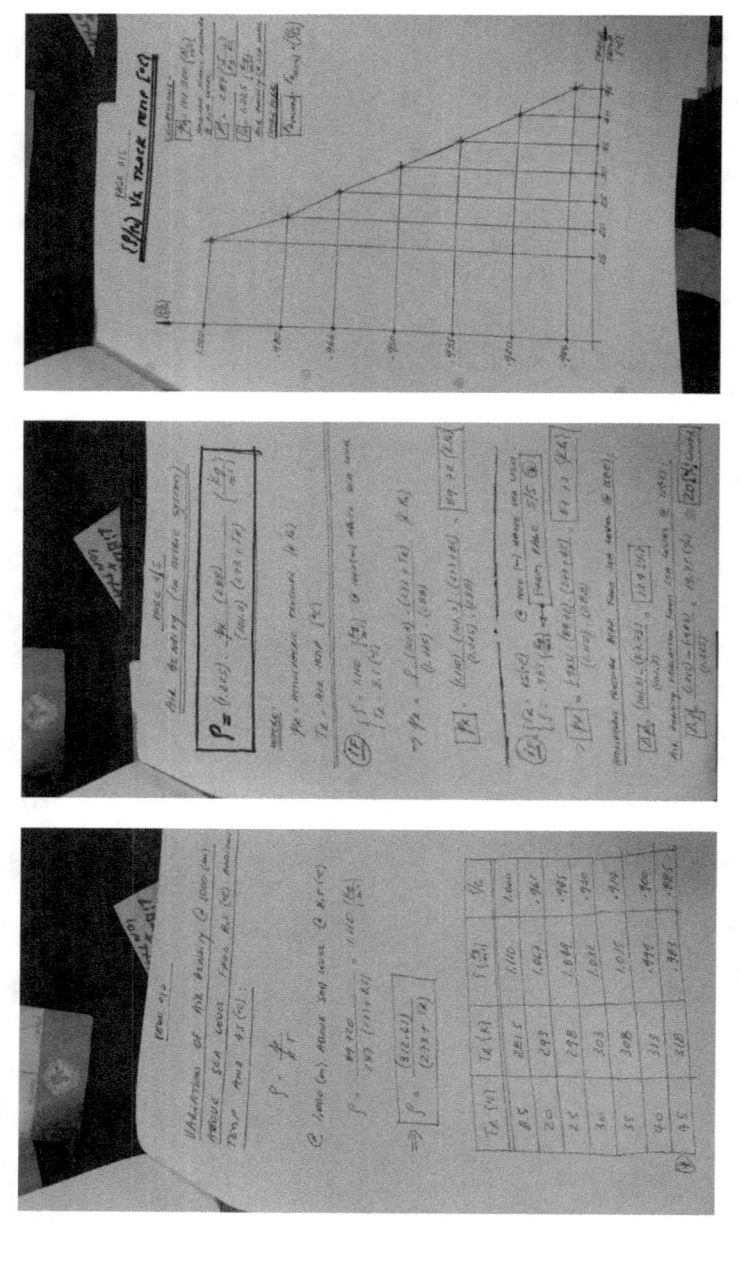

PAGE 124 — [P/h] Vs TRACK TEMP [°C]

PAGE 113

Variation of air density @ 1000 m above sea level from R.T.(C)

$\rho = \dfrac{P}{T}$

T_a (K)	T_a (C)	P (?)	%
85	281.5	1.110	1.000
20	293	1.061	0.961
25	298	1.044	0.941
30	303	1.030	0.930
35	308	1.015	0.910
40	313	0.998	0.900
50	323	0.953	0.863

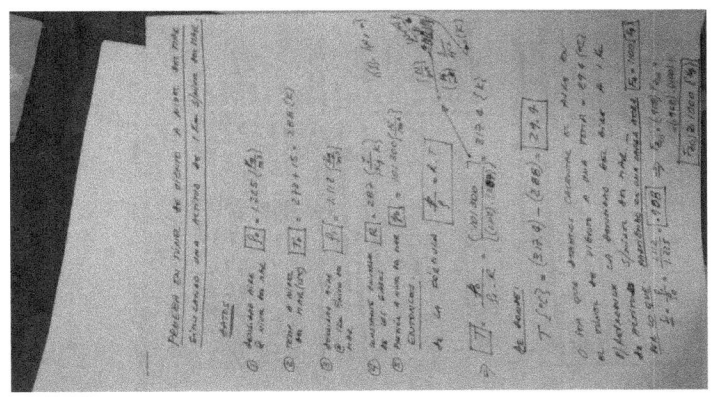

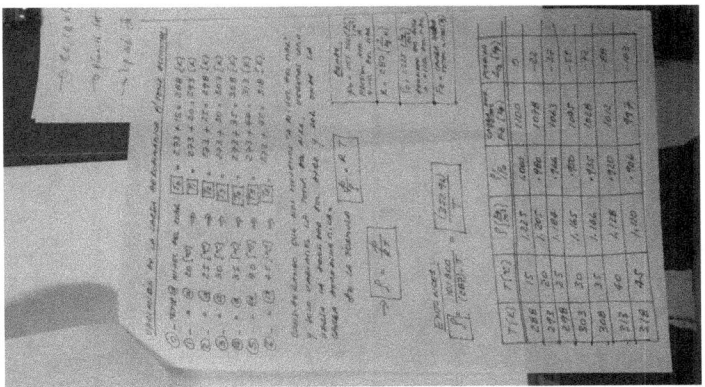

CONCLUSIONS

Along these 2 firsts chapters, we have covered the most important properties of air and its relations.

On the other hand, we have described what the boundary layer is and what the origin of turbulence is.

In the following chapters, we will see different methods to keep control of the boundary layer and its dimensions for a right CFD simulation. Likewise, we will analyze the dynamic consequences that the analyzed properties have.

In short, we want to lay the foundations to understand air principles in an easy and simple way.

Basic idea – reflexion:

- The friction between molecules is the same as to say viscosity.

- The friction air in the surface creates the boundary layer and the Coanda effect.

The next image, is very important in order to understand a lot properties: Viscosity, Coanda (also Bernouilli):

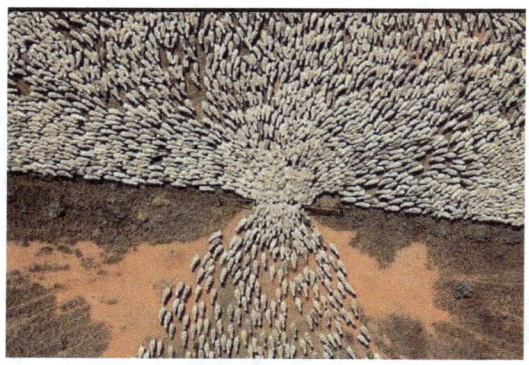

The sheep flock direction is up-down. AMAZING IMAGE ¡¡iiiiiii